Earthquakes
Science and Society

Earthquakes
Science and Society

David S. Brumbaugh
Northern Arizona University

PRENTICE HALL, Upper Saddle River, New Jersey 07458

Library of Congress Cataloging-in-Publication Data

Brumbaugh, David S.
    Earthquakes: science and society / David S. Brumbaugh
        p.   cm. — (Prentice hall earth science series)
    Includes bibliographical references and index.
    ISBN 0-13-523847-1
    1. Earthquakes. I. Title.   II. Series.
QE534.2.B78   1999
551.22 — dc21   99-30046
                                  CIP

To Michele and Mark.
Also in memory of Dottie, Scott, and Kenny.

Executive Editor: Pat Lynch
Production Editor/Page Layout: Kim Dellas
Manufacturing Manager: Trudy Pisciotti
Copy Editor: Maureen Mackey
Art Director: Jayne Conte
Cover Designer: Bruce Kenselaar

© 1999 by Prentice-Hall, Inc.
Simon & Schuster/A Viacom Company
Upper Saddle River, NJ 07458

Printed in the United States of America

10 9 8 7 6 5 4 3 2 1

ISBN 0-13-523847-1

Prentice-Hall International (UK) Limited, *London*
Prentice-Hall of Australia Pty. Limited, *Sydney*
Prentice-Hall Canada Inc., *Toronto*
Prentice-Hall Hispanoamericana, S.A., *Mexico City*
Prentice-Hall of India Private Limited, *New Delhi*
Prentice-Hall of Japan, Inc., *Tokyo*
Simon & Schuster Asia Pte. Ltd., *Singapore*
Editora Prentice-Hall do Brasil, Ltda., *Rio de Janeiro*

Contents

Preface

Can there be a topic more exciting than earthquakes? I don't think so. Whenever an earthquake large enough to cause significant damage and casualties occurs, it dominates the news, as well it should. In such instances it becomes clear that most people, including many in the media, have only a hazy understanding of earthquakes. This lack of understanding comes home all too strongly when one tries to teach about earthquakes at the university level, as I do.

For several years now, I have taught an introductory earthquakes course, primarily for nonscience students. It is quite a challenge, for although a number of videos and slide sets exist to supplement the spoken word, I cannot take my students on a field trip to *visit* an earthquake. Nor are simulated earthquakes easily available everywhere in the United States, as they are in southern California (e.g., Universal Studios). Much more frustrating to me, however, has been the lack of what I felt was a suitable textbook. From this need grew the urge to write this book.

Earthquakes: Science and Society was planned to be more than just an introductory college textbook, but a resource that would suit a variety of needs. The philosophy is: come to the table and take what you need, for there are certainly many reasons for wanting to learn more about earthquakes. Thus, the structure of this book was designed to actually represent three interconnected books that might stand alone. Part I: Earthquake Basics, says it all. This section provides the reader with the background on what an earthquake is, and how over the years scientists began to understand earthquakes. This required making the connection between earthquakes and faults as well as developing the necessary instruments and techniques to understand the earthquake process. A fascinating introduction to Part I is the development of human belief about earthquakes that incorporate early ideas and myths.

Part II: Earthquake Data Analysis and Its Contributions to Science, builds on the material of Part I. This section will appeal even more to the scientifically curious, although as with the rest of this book, no scientific background is presumed. Part II is applications. How do we use the data gathered from earthquake instruments? Nothing could be more fundamental as a result of studying earthquakes than determination of earthquake location and size. Neither location nor size is as easy to determine as one might assume from reading about earthquakes in the popular press. Certainly size can be confusing, as was learned by following reports after the 1989 World Series Earthquake south of San Francisco. A confused television reporter informed viewers that the magnitude of this event was 7.1 at the epicenter, but only 6.9 some distance away!

Seismologists have understood for many years that the movement of the ground recorded at a station hundreds or even thousands of miles away from an earthquake is literally loaded with information about what has happened at the site where the energy was released. Ground movement recorded at several stations is usually sufficient to determine location, size, fault plane orientation, direction and amount of fault slip, and how rupture proceeded with time. How this is done will be unravelled for the reader in Chapter 5.

The last two chapters of Part II outline two other triumphs resulting from the study of earthquakes. Chapter 6 shows how analyses of earthquakes from around the world provided convincing independent evidence in support of the plate tectonic theory. This chapter also provides a good synopsis of the development of the plate tectonic concept.

Chapter 7 shows how earthquake studies have revealed the hidden interior of Earth in great detail. This chapter follows the clues revealed from wave velocities and reflection and bending or refraction of wave paths. Truly, without earthquake data, we would know very little about Earth's interior.

Part III focuses on personal safety and earthquakes. The first lesson in Chapter 8 is what can be learned from past earthquakes. As the saying goes, if one does not learn history, one is doomed to repeat past mistakes. Beginning with the Kourion earthquake of 365 A.D. to the Mexico City tremor of 1985, these historic earthquakes are used as examples of the kinds of hazards related to earthquakes that one might expect and comprise a sort of what-not-to-do handbook.

Especially useful to American readers is a chapter on earthquake geography in the United States. It is well known that California is earthquake country. But what about tremors in Arizona? Washington state? What places in the United States are essentially devoid of earthquakes?

Of course it would be nice to avoid earthquakes altogether, or at least know when they will occur. Earthquake prediction, covered in Chapter 10, is a long-standing dream of the human race, which so far has only been achieved sporadically. Chapter 10 deals with past and present attempts at prediction, and the techniques available.

The last two chapters comprise the handbook for personal safety in the event of an earthquake. Chapter 11 outlines the steps to take before, during, and after an earthquake. The best thing to do prior to an earthquake is to choose a safe place to live. This information on siting is followed by tips on what makes a safe building, one resistant to earthquake damage or collapse.

Finally, appendices and a glossary supplement the body of the text. These include E-mail addresses of institutions and organizations that will be sources of more information on earthquakes.

For whom did I write this book? Everyone and anyone interested in earthquakes. No background is presupposed. Although it is aimed at first-year college students, I would expect it could easily be used in high school, or by the general public. All or part of the book could be used depending on one's interest or needs. For example, Part I serves primarily as an introduction to the development of the science of seismology and might attract those with interests in the history of science and the scientists of the past, or in mythology.

Part II is the scientific core of the book and deals solely with understanding earthquakes, the earthquake process, and how the study of earthquakes has been applied to understand Earth. This section contains information essential in any course taught about earthquakes.

Part III could appeal to sociologists, emergency response personnel, planners, and just about anyone interested in personal safety. This does not complete the list, but I have attempted in writing this book to make it versatile and to appeal to the widest possible audience.

I have been helped by many people in writing this book. I am grateful for the help and guidance provided by the editors at Prentice-Hall, Bob McConnin and Kim Dellas. Bob helped me get the project off the ground and nurtured it most of the way through. His advice and experience were invaluable. Kim helped put on the finishing touches and see the project through to the end.

I am indebted to the following reviewers for detailed critiques: Charles Ammon, Craig de Polo, Yildirim Dilek, Barbara Romanowicz, Charles Sammis, Don Steeples, Michael Stickney, Ivan Wong, and Terry Wright. Despite these reviews there will likely be mistakes. These are mine and mine alone, whether the result of errors of understanding or errors in expression.

I would certainly not have written this book without the inspiration of my teachers at Indiana University: Jud Mead who introduced me to the mysteries of geophysics and made it fun; Al Rudman whose excitement about earthquakes was contagious; and Tom Hendrix who led me to an appreciation of all rock structures, especially faults. I would like to add to this list a note of gratitude to Father William Stauder S.J. who took time out of a busy life to answer some questions a young graduate student had about earthquakes.

Finally I would like to thank my children Michele and Mark for their patience and understanding during the writing of this book. Their love and support was crucial.

PART I:

EARTHQUAKE BASICS

CHAPTER 1

Earthquakes: Myths, Legends, and Logic

MYTHS, LEGENDS, AND GODS: EARLY IDEAS ON THE ORIGIN OF EARTHQUAKES

Probably the last thing in the world to cross the mind of a person experiencing violent ground shaking during an earthquake is the cause of ground movement. Nevertheless, humans are a curious lot, and for as long as people have experienced earthquakes, they have wondered about their cause. Explanations of earthquakes have varied and usually bear a direct relation to the state of development of human society at the time.

An earthquake is just what the word suggests, movement of the ground. The movement can be due to several causes, which is one reason our ancestors often gave conflicting explanations for the occurrence of earthquakes. Early societies did not have the benefit of modern technology to gather information useful in understanding earthquakes. They were limited to simple observation of what was occurring at Earth's surface, and would combine this with guesses about Earth's interior, which was the obvious source of earthquakes. Often these observations also had to agree with cultural concepts prevalent at the time. Given all of this, it is not surprising that modern ideas about the causes of earthquakes go back only about 200 years.

It must be understood that many of the early causes suggested for earthquakes, and passed down to us as myths or legends, really result from a combination of simple observation, logical thought processes, and limited knowledge of the time. A paleolithic hunter, experiencing ground movement of an earthquake, would note the tremendous power required to move Earth's rocks. Power and motion in the mind of the hunter would be associated with some supernaturally large animal, similar to those in the environment. In Kamchatka, Russia, the native people believed that earthquakes occurred when a mighty dog, Kozei, shook freshly fallen snow from his coat. Usually the giant animal was placed underground, where it could not be seen, and where observation suggested the earthquake originated. Often, the entire Earth would rest on the back of an animal, also conveniently not visible. The Chinese claimed that Earth rode on the back of a giant ox, while the Mongolians believed it was a frog (Fig. 1.1). Whenever the animal moved, Earth shook and trembled.

The ancient Japanese believed that a giant catfish was responsible for earthquakes; the earthquake namazu (Fig. 1.2). The namazu was so prone to mischief that it had to be controlled by a god. If the god looked away, or left for a ceremony, the namazu would flop about, causing a quake. A catfish may seem a curious choice. However, observers have reported what they thought was unusual behavior in catfish before earthquakes. Perhaps the ancient Japanese made similar observations.

India has a long history of devastating earthquakes, and because of abundant wildlife that formed an important part of rural life, animals were incorporated into myths as a cause of the quakes. One belief held that seven serpents were the guardians of the seven sections of the lowest heaven. They also took turns holding up Earth. When one finished and another moved in place to take over, people on Earth felt the ground move and shake. India has a history of great cultural diversity, and so it is not surprising that there were a number of competing beliefs about the causes of earthquakes. Another Indian belief on the cause of earthquakes maintained that Earth was held up by four elephants that stood on the back of a giant turtle. The turtle in turn stood on a cobra, all of which makes for a shaky foundation. When any of these animals moved, Earth trembled.

The advantage of using supernaturally large animals or giants is that they are sources of power equal to the magnitude of an earthquake. Of course, no one is able to see any

Figure 1.1   The Mongolian belief about the cause of earthquakes was that it was due to movements of a large frog, which supported the world on its back.

Figure 1.2   In Japanese folklore, earthquakes were caused by a giant catfish
(namazu) beneath the ground. When the catfish flailed about, the ground shook.

of these creatures, which are conveniently hidden from view. A second approach of combining supernatural power with an invisible source was to blame earthquakes on a deity.

The god Loki in Scandinavia was held responsible for the trembling ground. Loki was tied to a rock in a cave underground as punishment for killing his brother. A serpent would drip poison down on him, which was caught in a bowl by Loki's sister. When she emptied the bowl, Loki had to twist and turn to avoid the poison, thus causing earthquakes.

The Greeks had a pantheon of gods to choose from in explaining earthquakes, many of them abiding on Mt. Olympus. The greatest of these was Zeus. As in Prometheus's speech to Hermes:

> There is no force which can compel my speech.
> So let Zeus hurl his blazing bolts,
> And with the white wings of the snow,
> With thunder and with earthquake
> Confound the reeling world.

The Greeks frequently appealed to the power and capricious behavior of the gods in order to explain earthquakes. They considered that the temper tantrums of the god of the sea, Poseidon, was the cause of earthquakes. For example, it is said that one time Poseidon was angered by the giant Polybites, and chased him across the Aegean Sea, setting off earthquakes wherever his feet were put down. Poseidon was known to the Greeks not only as god of the sea, but was frequently referred to as Earth shaker. The trident Poseidon carried was believed to cause earthquakes when he struck Earth with it (Fig. 1.3).

The inhabitants of Mexico also blamed the gods for earthquakes. Mexico is a land close to the cataclysmic events of nature, especially earthquakes and volcanic eruptions. The 1985 earthquake killed an estimated 10,000 people in Mexico City, and this same skyline is dominated by the cones of two volcanoes with Aztec names: Popocateptl and Iztaccihuatl.

The region around and including Mexico City was the homeland of the Aztec Empire 500 years ago. Mindful of their environment the Aztecs incorporated volcanic eruptions and earthquakes in their theology. When the Aztecs strayed from the gods and became wicked, it was predicted that the gods would destroy them by massive quakes. Such events were depicted on the Aztec calendar (Fig. 1.4). Within the Aztec culture their calendar summarized their history, and their fate. At the center of the calendar

Figure 1.3   Poseidon the Earth shaker. Vibrations created by Poseidon striking the ground with his trident were thought to cause earthquakes.

Figure 1.4 The Aztec view on the cause of earthquakes was an integral part of their culture and history. The center part of the stone Aztec calendar shown here predicts destruction by massive earthquakes. In the rectangles of the earthquake sign are the dates at which the end of an age occurs.

was the image of the Sun god. Ringing the Sun were the images of various ages that ended with earthquakes (rectangles). The last age was expected to come to an end by the occurrence of massive earthquakes.

The native peoples to the north in what is now the United States also had some interesting stories about deities and earthquakes. The Chickasaws believed that at least one earthquake was related to the troubles of a young chief named Reelfoot. He was thus named because of a twisted foot. Reelfoot was in love with a Choctaw princess but could not marry her because of her father's refusal. The chief and his warriors kidnapped her and began to celebrate their marriage. The Great Spirit became angry at this behavior and stomped his foot. The ground shook and caused the Mississippi to overflow its banks and drown the entire wedding party. This caused Reelfoot Lake to form, which stretches along the western boundary of Tennessee. This lake actually was formed during a sequence of earthquakes during 1811 and 1812.

The threads running through all of these beliefs are power, and powerlessness, and often punishment as well. Only a god could have the awesome power at his command to visit an earthquake upon powerless humans as a payment for their sins. For example, one Byzantine emperor prohibited swearing, blasphemy, and public kissing to alleviate the threat of earthquakes from God.

Earthquakes were often frequently associated with significant theological events. Perhaps the best known would be the earthquake reported by St. Matthew at the time of Christ's death: "and the earth did quake, and the rocks rent" (St. Matthew 27:51). Almost as well known might be the earthquake that led to the release of Paul and Silas from jail in Macedonia: "And suddenly there was a great earthquake, so that the foundations of the prison were shaken: and immediately all the doors were opened, and everyone's bands were loosed" (Acts 16:26).

Even early scientists gave some credence to divine intervention in the matter of earthquakes. A Belgian chemist in the seventeenth century explained the system of earthquake occurrence as due to God signaling an angel to strike a bell. The vibrations from the bell shook the atmosphere, which in turn vibrated the ground. Indeed, as late as 1752 the British Royal Society, the equivalent of the National Academy of Science in the United States, declared that earthquakes could occur only where people needed chastening.

Earthquakes in Italy in 1570 sparked interest in the phenomenon among writers of the time. Galesius wrote a treatise on earthquakes that summarized causes of quakes going as far back as the Greeks. He also gave a detailed analysis of earthquake type based on ground motion (e.g., upward movement). Even more interesting is Galesius's discussion of the distribution of earthquake regions. This was one of the earliest attempts to describe earthquake geography and to recognize that some areas were prone to earthquakes, while others were free of them.

THE NATURAL WORLD AND EARTHQUAKES

Early societies usually had widely accepted ideas about the nature of Earth and the universe that were also incorporated into beliefs about earthquakes. For example, the ancient Greeks believed everything was composed of four basic elements: earth, fire, water, and air. It was a natural step to ascribe the causes of earthquakes somehow to these elements.

Anaxogoras (500–428 B.C.) believed that the collision of gases in caverns generated fire. As the fire rose it burst through obstacles violently, causing the ground to shake. Thales (c. 580 B.C.) held that Earth was like a ship floating on water and that the movement of the water caused earthquakes. Archelaus (fifth century B.C.) appealed to compressed air in caverns. Air entering caverns from the surface eventually became compressed, causing violent storms to shake Earth. Such ideas could be related to simple observation of air rushing out of the ground. For example, at Wupatki National Monument in Arizona, air in and outside of channels formed by rock fractures are sensitive to barometric and temperature changes. At regular intervals the air will be drawn into and forced out of these channels.

Aristotle believed gas played an important role in the occurrence of earthquakes. However, he also noted a connection between volcanic eruptions and earthquakes. When violent eruptions occurred, the ground would quake and tremble. Because volcanic eruptions often included great and violent releases of gas, it was logical to suppose that the movement of such gases from cavern to cavern before the eruption shook the ground, causing earthquakes.

Such ideas about the causes of earthquakes were passed down to succeeding generations and to other cultures, changing somewhat as they did so. The Romans borrowed much in architecture and theology as well as in science from the Greeks. The Roman writer Seneca in his *Questiones Naturales* believed as Archelaus did, that confined subterranean air was the principal cause of earthquakes. Although erroneous, such an association might be understandable to someone in a house when the very walls shake from strong gusts of wind. Indeed, in some parts of the world, building codes have more demanding requirements for wind resistance in building structures than for earthquakes.

The idea of vapors as a cause of earthquakes was a very persistent one, lasting for centuries from the time of the ancient Greeks until at least the 1700s. It was also frequently combined with a humanistic concept for Earth. This merging of human characteristics with Earth began in classical times and continued well into the Renaissance and Shakespeare's time. In this view, Earth possessed a circulatory system, much like a human being. Blood and bodily fluids were compared to underground rivers. When this system became upset or disturbed, Earth became sick. It was this view that Shakespeare had in mind when he wrote:

> Diseased Nature oftentimes breaks forth
> In strange eruptions: oft the teeming earth
> Is with a kind of colic pinch'd and vex'd
> By the imprisoning of unruly wind
> Within her womb which for enlargement striving
> Shakes the old beldam earth and topples down
> Steeples and moss grown towers."
> *King Henry IV*, pt.1, act 3, scene 1

This is an interesting view inasmuch as it suggests a rudimentary attempt to apply knowledge of the behavior of gaseous materials. In the case of Shakespeare's quotation, the allusion is made to the natural ability of gas to expand and exert pressure when confined, resulting in earthquakes. Indeed, as late as 1758 Isnard suggested relieving pressure and preventing earthquakes by digging deep shafts into Earth.

THE AGE OF REASON: THE EIGHTEENTH CENTURY

By the late Middle Ages the roots of modern science had begun to grow. The wisdom of the ancients was no longer accepted without question, and the restrictions placed on science by theology had diminished. The concept that God does not interfere in the affairs of nature, but works through nature, was an important new viewpoint. This allowed students of science to study nature without concerning themselves with theological questions or causes.

The final link in the development of modern science was the evaluation of hypotheses by experimentation. Only through experimentation could information be acquired to prove or disprove ideas based on observation and theory. The advances built on this base during the sixteenth and seventeenth centuries were astonishing. The crowning achievements of Newton in theoretical mechanics/gravity and Galileo's experiments in gravitational attraction are well-known examples.

The development of modern seismology was stimulated by the Lisbon earthquake of November 1, 1755, an event that stunned Europe, leveling one of Europe's capital cities. Among the people who studied and wrote about the Lisbon earthquake was an astronomer in the modern scientific tradition, John Michell. A man of many interests, Michell was the inventor of the torsion balance used by Henry Cavendish to measure the universal gravitational constant.

Although Michell did not make a great contribution in understanding the cause of earthquakes (he ascribed it to subterranean fire, producing water vapor, sort of like a boiler exploding), he did make a number of contributions resulting from his study of the Lisbon earthquake. He was the first to associate earthquakes and faults (Fig. 1.5), although it is not clear if Michell grasped the nature of the relation. Just as importantly, however, Michell did understand that the ground shaking in earthquakes was related to the passage of waves through the rocks. He proposed an analogy:

> Suppose a large cloth, or carpet (spread upon a floor), to be raised at one
> edge, and then suddenly brought down again to the floor; the air under it,
> being by this means propelled, will pass along, till it escapes at the opposite
> side, raising the cloth in a wave all the way as it goes.

Unfortunately, as a cause of the waves Michell imagined vapors to raise the strata of Earth in a wavelike fashion during the earthquake (see Box 1.1).

Once having recognized the passage of waves in the ground as the immediate cause of ground shaking, this led to two other contributions by Michell, which followed naturally. Traveling waves propagate at a velocity that can be estimated. Using simple de-

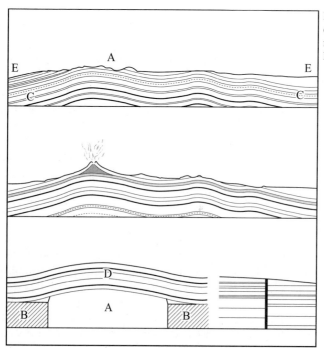

Figure 1.5   Earthquakes were caused by a disruption of geologic rock layers, according to John Michell.

BOX 1.1

Waves in Nature

Wave motion is a very common occurrence in nature. One example would be waves in water, with the familiar to-and-fro motion. Wave motion results from the transfer of energy. A rope tied at one end and disturbed by a sharp snap of the wrist at the other end creates a wave. The wavelike motion travels down the length of the rope away from the wrist, while the particles in the rope move only up and down, and come to rest where they started.

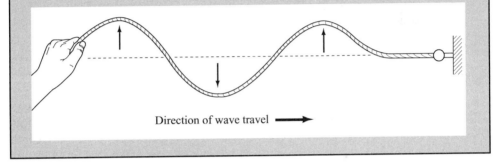

Direction of wave travel ⟶

---

duction based on observer reports of the Lisbon earthquake, Michell estimated the earthquake wave velocity as 0.5 kilometers per second. This was the first estimate of earthquake wave velocity, and, as we shall see, it is rather low, unless he was oberving velocity in loose, dry soil, but not really so bad considering Michell's method of measuring the travel time of the waves. This is a very important contribution because the velocity of earthquake waves is central to understanding earthquakes and Earth's interior.

Acceptance of the concept of earthquake waves is also a necessary first step leading to earthquake location. Because waves spread out and travel in specific directions, much like the waves in a pond spreading out from the point where a stone was tossed in the water, Michell imagined it would be possible to draw lines in these directions, and that the lines would intersect at the geographic location of the earthquake. Michell also noted that the interval between the time at which the earthquake occurred and the time of arrival at different points could be used to locate an earthquake. However, it was a long time (150 years) before any of these ideas could be translated into a method useful in locating earthquakes.

Recognition that earthquakes originated at a specific location naturally led to the question of what was occurring at the earthquake source. An important and necessary step in the development of the modern science of seismology was the recognition of *faulting* as the principal cause of earthquakes. This would not be an easy step to achieve as most earthquakes do not cause faulting at Earth's surface. This relation was further obscured in people's minds by the fact that ground shaking can also result from volcanic eruptions. Also, as previously indicated, ground shaking can result from strong wind or wave action against a shoreline as well as from other causes. But clearly, when all of these external causes were absent, and in areas of no volcanic activity, the ground shaking was a mystery whose origins could only be the subject of guesswork.

From the earliest times, those who studied rocks at Earth's surface recognized that mountainous areas were usually composed of folded, bent, and fractured or faulted rocks. During the nineteenth century, an idea arose that was offered to explain the distorted rocks in mountain systems. This was the Earth Contraction Hypothesis. Basically, this suggested that, over time, Earth was cooling and shrinking. As Earth shrank, the outer layer suffered compression parallel to the surface. The pressure caused rock to bend and break.

Robert Mallet was an Irish engineer with an interest in earthquakes. As early as 1850 he was experimenting with explosives to measure the velocity of elastic waves. His study of the 1857 Neapolitan earthquake resulted in a report that helped lay the foundations of modern observational seismology. Although he emphasized the role of volcanoes as the cause of earthquakes, Mallet also implied faulting could be involved. Invoking the contraction theory, Mallet (1859) stated that earthquakes could be caused by "the breaking up and grinding over each other of rocky beds."

A geologist of considerable influence and stature in the late nineteenth century, Edward Suess (1875) made the first clear statement of association: "Earthquakes occur along the lines of tectonic movement in a mountain system." Neither Suess nor Mallet had actually observed fault slip during an earthquake, but they were formulating hypotheses based on deduction from observation of faults.

The relationship between faults and earthquakes gained support in the 1880s through the work of an American geologist, G. K. Gilbert. Gilbert recognized through his work and the work of others that faults and earthquakes were related. Gilbert had visited fresh fault scarps of the 1872 Owens Valley earthquake. That he understood that a relationship existed between faults and earthquakes is seen from an article published in the September 20, 1883, issue of the *Salt Lake Tribune*. Gilbert stated that "when an earthquake occurs a part of the foot slope goes up with the mountain, and another part goes down (relatively) with the valley. It is thus divided and a little cliff marks the line of division.... This little cliff is, in geologic parlance, a fault scarp."

Clear understanding of the relationship between faults and earthquakes would come from the actual observation of a fault scarp forming at the time of an earthquake. The critical event was an earthquake in 1891 in Japan. This was the Mino-Owari earthquake, named after the prefectures (political regions) in which the most severe ground shaking occurred. Observers noted that a fault scarp had formed when the tremor occurred. The fault scarp was over 100 kilometers long and up to 5 meters (18 feet) high, cutting and offsetting roads (Fig. 1.6). The fault scarp was studied by Bunjiro Koto, professor of geology at the Imperial University. In a report resulting from this study, Koto clearly established the correct order of cause and effect:

> The sudden elevations, depressions, or lateral shiftings of large tracts of country which take place at the time of destructive earthquakes are usually considered as the effects rather than the cause of subterranean convulsions; but in my opinion, it can be confidently asserted that the sudden formation of the "great fault of Nes" was the actual cause of the great earthquake.
>
> (Koto, 1893, *College Science Journal.*, no. 5, p. 352)

Figure 1.6   The Mino-Owari fault scarp that resulted from the 1891 Japanese earthquake. Notice the displacement of the road crossing the fault scarp.

The next step in understanding earthquakes was to recognize how earthquakes resulted from faulting. This would require more data, which could only come from another large earthquake and the surface rupturing fault. The earthquake that would provide the necessary observations would turn out to be the 1906 tremor that shook San Francisco.

THE ELASTIC REBOUND THEORY

The April 18, 1906, San Francisco earthquake tore the earth in a 430-kilometer-long gash, offsetting roads and fences and causing locally severe damage to structures. One of the largest earthquakes to strike the United States in historic time, and causing extensive damage to the city of San Francisco, this event had a tremendous impact on the public. One outcome of public concern was the establishment of a commission to study the tremor. People wanted answers. Why had the earthquake occurred? Was such an event likely to happen again? If so, when? These were some of the questions requiring answers that placed pressure on the members of the commission. Fortunately, the quake had provided a lot of clues, not the least of which was the long and prominent gash in Earth's surface. At places along the tear, the two sides had moved past one another as much as 21 feet. The region to the east of the fault had moved in a southeasterly direction compared to the ground on the west side, which showed a relative shift toward the northwest.

The fault rupture crossed the heart of a ranch in Marin County, north of San Francisco, where structures displaced by the movement on the fault showed a relative horizontal shift of 15.5 feet (Fig. 1.7). This sort of data was abundant and essential to understanding the cause-and-effect relation of faults and earthquakes. The early activity of the commission consisted of seeing that all available data on damage, observer reports, and fault displacement were collected and assembled.

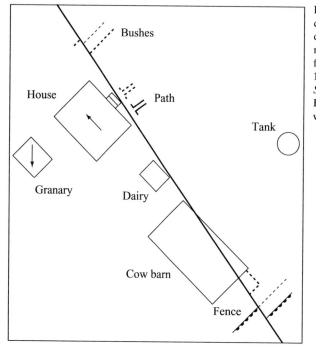

Figure 1.7 Right lateral displacement of about 15 feet as documented at the Skinner ranch north of San Francisco resulting from fault slip associated with the 1906 earthquake. (From *Elementary Seismology* by Richter, ©1958 by W. H. Freeman and Company. Used with permission.)

It was clear that the fault had released a great deal of energy in the 1906 earthquake when it ruptured and moved. But from where had the energy come? The commission chairman was a prominent scientist, H. F. Reid, who realized that other data existed that might shed some light on earthquake processes.

If objects such as fences and roads could document displacement due to fault movement, what had the countryside been like before the earthquake? Fortunately, California as a growing frontier state had numerous surveying lines established across the fault in previous years, to lay out roadways, city boundaries, and property corners. Detailed surveys covering large areas were carried out in 1851 to 1865 and 1874 to 1892. Another survey was conducted just after the quake for comparison purposes with the earlier surveying data. The results were surprising. It was clear that movement occurred along the fault at the time of the 1906 earthquake. What was not suspected was that the ground close to the fault had been moving for 50 years before the 1906 quake! But the fault was not moving at the time the land around it was. This led to the concept of *elastic rebound*, as outlined by Reid in the final report of the earthquake commission.

Although it may not seem like it, rocks, like many other natural and manufactured materials, are elastic. There is a great difference in the elasticity of a rock and a rubber band, but the concept is the same. If we stretch a rubber band, its increase in length is an elastic response. This is especially so because the stretching is not permanent. When we release the rubber band and measure its length, it is the same length as before it was stretched. Thus, elastic deformation is recoverable, not permanent. Stronger materials when thin are also visibly elastic. Sit on the metal hood of an automobile and it depresses. When you get up it recovers its original shape. However, every elastic material has an elastic limit, the point beyond which further deformation is no longer recoverable, but permanent. If one car hits another for a fraction of a second the response may be elastic. Unfortunately, deformation usually continues as force is applied. The metal will no longer behave elastically under the increased force and it crumples, deforming permanently. Metal is not very elastic. But a simple experiment shows that rocks, while also not very elastic, do have elastic properties. Experiments in the lab with a press and a thin layer of rock demonstrate rock elasticity (Fig. 1.8).

As long as force is continuously applied to an elastic material, that material will deform, storing energy. An example would be a wood pencil. Bend it between the thumb and forefinger. As long as it remains bent, the pencil has stored energy known as *potential energy*. As the term suggests, the stored energy can be released. If the bending action is continued the pencil breaks and the potential energy is released as energy of motion, or kinetic energy (Box 1.2). The snapping sound heard when the pencil breaks is the result of the energy release. The energy travels away from the point of release as a sound wave, reaching our eardrums.

H. F. Reid suggested that the rocks on both sides of the San Andreas fault were being elastically deformed long before the 1906 San Francisco earthquake. As the rocks bent under force, and changed position, they were storing strain energy (Fig. 1.9). This could continue until the strength of the rocks was exceeded and the rock mass would give way at its weakest point, the fault surface. The data analysis further suggested that upon release of the energy, the rocks around and on both sides of the fault snapped back or rebounded elastically. The elastic rebound theory fit the facts gathered from the 1906 San Francisco earthquake and today remains a generally accepted description of the cause-and-effect relationship between faulting and earthquakes.

SUMMARY

Prehistoric explanations of earthquakes coming down to us in the form of legends emphasize the role of giant animals to account for the powerful forces associated with earthquakes. Usually such animals were conveniently underground or out of sight. An equally powerful source invoked to explain earthquakes was the capricious or wrathful

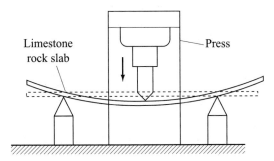

Figure 1.8   A simple lab experiment with a rock press and a thin slab of rock demonstrates the elastic properties of rock.

BOX 1.2

Behavior of Materials

The behavior of material may be described in terms of response to the application of force. The response can be seen as a change in shape (e.g., length) or volume. Scientists testing the behavior of materials do so by compressing samples in a mechanical press between two anvils and recording the results. A granite rock sample will first respond elastically by shortening. With a release of force, the sample springs back (elastically) to its original length. The deformation is recoverable. But every material has an elastic limit, which is a limiting force beyond which deformation becomes permanent. In the case of the granite under the conditions just described, this would be fracturing of the sample.

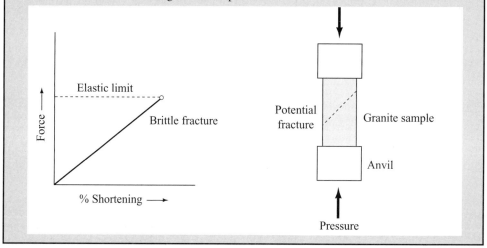

behavior of a god or gods. Acceptance of divine intervention as a cause of tremors extends from the Greeks and, probably before, down to modern times, and is still believed by some people today.

Early empirical explanations appealing to physical phenomena were based on incomplete data. The classical Greeks invoked the action of subterranean vapors trapped in caverns. This idea proved popular and, with minor variations and additions, became passed down to succeeding generations, with adherents on record as late as 1758.

The development of the modern scientific method beginning in the late medieval period prepared the way for the modern study of earthquakes. Experimentation gradually emerged as an important way to test ideas based on observations. A new generation of scientific workers experienced in the scientific method was ready when the great Lisbon earthquake of 1755 occurred. An astronomer, John Michell, made the first study of earthquake waves, attempting an estimate of their velocities and suggesting techniques that might be used for location of the source.

But the earthquake source was not to be clearly identified and understood until the late nineteenth century. The formation of a fault scarp was observed as a result of the

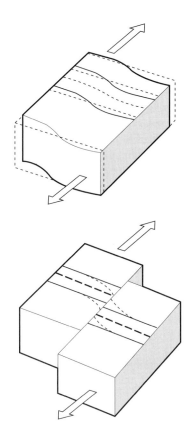

Figure 1.9   As force builds up in a rock, the rock responds elastically by bending. Eventually enough force is built up in the rock that friction on the fault surface is overcome and the fault slips, releasing energy as earthquake waves. A highway serves as a fault slip reference marker on the tops of the blocks.

1891 Mino-Owari earthquake and the correct cause-and-effect relation established by B. Koto. The 1906 San Francisco earthquake provided additional important data. The report on this earthquake by the Reid Commission resulted in the elastic rebound theory, which explains the ground deformation around and across a fault before, during, and after an earthquake.

KEY WORDS

contraction theory	Mallet	vapors
culture	Michell	volcanoes
earthquakes	Mino-Owari	waves
elastic rebound	myths	
elastic waves	namazu	
experimentation	Poseidon	
faulting	punishment	
Galesius	Reid	
God	San Francisco	
Koto	Seneca	
Lisbon (1755)	Shakespeare	

C H A P T E R   2

Measuring Earthquakes

INTENSITY OF GROUND SHAKING

One of the fundamental pieces of data that can be collected after an earthquake is an observer's description of the ground shaking. The human body is a sensitive earthquake sensing tool and is able to detect a wide range of ground movements. Thus, it can be said that the first earthquake instrument available was the human body. This was fortunate, for as we shall see it took a rather long time to develop instruments to sense and record earthquake ground motions.

Another kind of data available in addition to observers' reports from earthquakes are damages caused to structures or disturbance of objects. This includes effects ranging from cracked walls and swinging doors on up to building collapse. Naturally, these effects can vary with the kind and type of construction or the objects involved, as well as with the size of and distance from the earthquake, the nature of the rupture, and local geological conditions. Effects on structures or on objects and observer reports of ground shaking taken together form the basis for the earliest scientific studies of earthquakes.

The first step in utilizing observer reports of ground shaking and damage information was to devise a scale that would arrange these reports as to level of intensity. The first attempt was by Domenico Pignataro in 1783, who devised an intensity scale with five levels: slight, moderate, strong, very strong, and violent. He used these intensities to describe over a thousand shocks of the Calabria, Italy, earthquake sequence of 1783 to 1786.

A more detailed intensity scale was devised by P. N. C. Egen in 1828. This scale had six levels, with *1* as lowest and with each level based on a more detailed description of events:

1. Only very slight traces of the earthquake can be sensed.
2. Few people, under favorable conditions, feel the shock; glassware close togeth er jiggle; small plants in pots vibrate; hanging bells do not ring.
3. Windows rattle; house bells are rung; most people feel the tremor.
4. Slight movement of furniture; shock in general is so strong it is felt by everyone.
5. Furniture shaken strongly; walls are cracked; only a few chimneys come down; the damage being caused is insignificant.
6. Furniture shaken strongly; mirrors, glass, and china vessels broken; chimneys come down; walls are cracked or overthrown. (See Box 2.1)

These intensity-scale criteria were derived from the results of the Netherlands earthquake of February 23, 1828, and placed on a map at the appropriate locations. This

---

BOX 2.1

Intensity Reports

Standardized earthquake report forms have been developed so that sets of intensity data collected for an earthquake are consistent. The results of these report forms are then tabulated and compared to the Modified Mercalli (MM) scale to assign an intensity level for each location. For example, the results from the form below would be assigned a level MM V intensity.

ARIZONA EARTHQUAKE INFORMATION CENTER
EARTHQUAKE FELT REPORT

Please complete this questionnaire and return as soon as possible. Please feel free to make additional copies for others who felt the earthquake. As every piece of information given by you is of value, please return the questionnaire even if only a few effects were notable. Thank you.

Date and approximate time of earthquake: 4/29/89
Name of person filling out form: STEVENS
Address:
City: ANGEL   County: COCONINO
State: AZ   ZipCode:

PERSONAL REPORT

Your location at time of earthquake, if different than above, was:
Address SAME
City   County
State   ZipCode
Did you personally feel the earthquake?   [X] yes   [] no
Were you awakened by the earthquake?   [X] yes   [] no
Were you frightened by the earthquake?   [X] yes   [] no
Were you at   [X] home [] work [X] inside [] outside   Other
Ground conditions: [] dry   [] moist [] wet/saturated
Ground type: [] sandy [] clay   [] fill dirt   [] hard rock
[] alluvium
Your activity when earthquake first occurred: [] walking
[X] sleeping   [] lying down   [] standing   [] driving
[] sitting   Other
Did you have difficulty in   [] standing up   [] walking   [] maintaining balance?
Vibration could be described as: [] light [] moderate   [X] strong
Type of vibration felt: [] sudden sharp jolt [] rolling
[] rocking   Other
Duration of shaking: [X] less than 10 seconds   [] 10-30 seconds
[] 30-60 seconds
Was there earth noise?   [] no   [X] faint [] moderate   [] strong
Describe noise: i.e. sounded like   THUNDER
Can you estimate the direction the noise came from? (mark any direction(s) that applies)
[X] north [] south [] east   [] west
Have you been through an earthquake before?   [] yes   [X] no
If yes, what state/country were you in?

COMMUNITY REPORT

Approximate population: [] less than 1,000   [X] 1,000-10,000   [] 10,000-100,000
[] over 100,000   [] rural area
The earthquake was felt by: [] few   [] several [X] many   [] all
The earthquake awakened: [] few   [X] several [] many   [] all
The earthquake frightened: [X] few   [] several [] many   [] all
What indoor physical effects were noted in your community?
Rattled: [] N/A [X] windows   [] doors [X] dishes;
[] slightly   [] loudly
Walls/floors creaked: [] N/A   [X] slightly   [] moderately
[] loudly

Building trembled(shook):   [] N/A   [X] slightly   [] moderately
[] strongly
Hanging pictures(more than one):   [] N/A   [X] swung [] out of place/tilted   [] fell
Water in small containers: [] not disturbed   [X] slightly disturbed [] spilled
Windows: [] N/A [] few cracked   [X] some broken out [] many broken out
Small objects overturned: [X] unmoved   [] shifted [] overturned
Glassware broken? [] dishes [] knicknacks;   [X] none   [] few   [] several [] many
Were   [] light furniture or [] small appliances; [X] not disturbed   [] displaced
[] overturned
Were   [] heavy furniture or   [] heavy appliances;   [X] not disturbed
[] displaced   [] overturned
Items thrown from shelves: [X] none   [] few   [] several [] many
Estimated direction items thrown: (mark any that apply)   [X] N/A
[] north [] south [] east   [] west
What outdoor physical effects were noted in your community?
People ran out of buildings: [X] none   [] few   [] many
Trees and bushes shaken: [X] none   [] slightly   [] moderately
[] strongly
Felt in stopped vehicles: [] none   [] slightly   [] moderately
[] strongly
Felt in moving vehicles: [] none   [] slightly   [] moderately
[] strongly
Water splashed on bank/sides of: [] lakes   [] ponds [] swimming pools
Other
Masonry fences/retaining walls were: [X] undamaged   [] cracked
[] broken   [] destroyed/fallen
Underground pipes: [X] undamaged   [] cracked   [] broken
[] out of service
Damaged: [X] none   [] sidewalks   [] streets [] highways
Describe damage:
Check below any structural damage to buildings.
Interior walls: [] N/A   [X] plaster [] dry wall   [] wallboards   [] no damage
[] hairline cracks   [X] few large cracks [] many large cracks
Exterior walls: [] N/A   [] brick   [] concrete block   [] stone   [] stucco
[] wood frame   [] brick veneer   Other   ;
[] no damage   [] hairline cracks   [] large cracks
[] bulged outward   [] partial collapse   [] total collapse
Other
Foundations: [] not disturbed   [] cracked   [] destroyed
Chimneys: [] none   [] few   [] several [] many;
[] undamaged   [] cracked   [] twisted [] bricks fallen
[] broken off roof line   [] fallen
What type of construction was the building that showed this damage? [] wood [] stone
[] wood frame w/ brick veneer   [] steel frame   [] concrete block
[] reinforced concrete   [] mobile home
Other

---

was the first attempt to relate ground-shaking intensity to geographic distribution. The limitation of this early scale was that it was designed for a moderate-sized earthquake and so could not be used for larger events. In fact, the early scales were generally only used by the person who devised them.

The next step in using ground-shaking intensity data was to define areas of equal intensity from the map of intensity numbers. The earliest known attempt to do this was by Johann Noggerath in 1847 for the July 29, 1846, Rhenish earthquake (Fig. 2.1). The map was divided into two areas of equal intensity of ground shaking bounded by lines

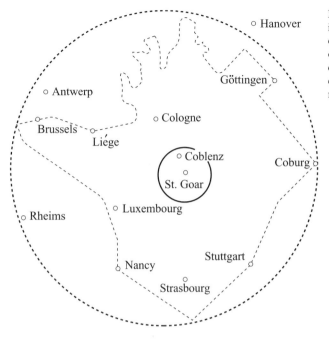

Figure 2.1   Ground-shaking intensity map for the Rhenish earthquake of July 29, 1846. The outer dotted circle is the limit of earthquake effects. The map is the earliest known isoseismal intensity map.

known as isoseismals (i.e., contours of equal shaking intensity). The innermost area, including Coblenz and St. Gaor, contained the highest intensity reports and was the likely location on Earth's surface above where the earthquake occurred at depth within the crust. Thus, a map defining areas of differing levels of ground-shaking intensities became a tool to locate a quake. The point of initiation of the earthquake within Earth is termed the *focus*, or *hypocenter*, while the point on Earth's surface or on a map directly above the focus is the *epicenter* (Fig. 2.1).

The highest intensity of ground shaking also became a way of measuring the relative size of earthquakes. In general, all else being equal, the higher the maximum intensity, the larger the earthquake. There are problems with using this kind of approach to estimate earthquake size, but for about a hundred years this was the only method available and in general use.

The intensity of ground shaking is determined by the manner and amount of energy released at the source, the distance from the source, including the depth of the source, and the material and topography at the site experiencing ground movement. Only distance was really well understood by early intensity mappers. It is clear that, as with most wave phenomena, a portion of the wave energy is absorbed by the material through which it passes. This is generally referred to as *attenuation of the wave*. It results in a decrease in wave height or amplitude and therefore in ground movement or intensity as well (Fig. 2.2). The farther the wave travels, the more energy is lost by absorption. In addition, spreading of the elastic wave front, in a way similar to the ripples expanding out from a stone thrown into a pond, causes a loss in wave amplitude and ground shaking intensity. The combined effect of both attenuation and geometrical spreading of the wave front means that the farther the site experiencing ground shaking is from the

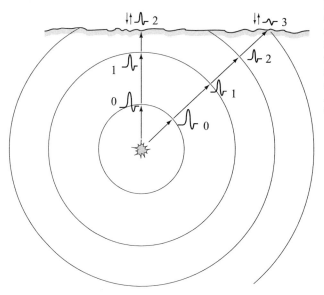

Figure 2.2 Loss of amplitude of earthquake waves with distance from the source. This has a direct effect on the intensity of ground shaking. The ground shaking (arrows at the surface) is greatest directly above the source. Amplitude decay also depends on wave frequency (see Boxes 2.6, 2.7).

source, generally the less intense is the ground shaking. Distance, of course, must also be considered in three dimensions, which means that the depth of the earthquake focus must be considered. As would be expected, deeper earthquakes tend to cause lower surface intensities than shallower ones of the same size.

The simple picture of ground-shaking intensity versus distance is complicated by source details and the rock type undergoing movement. Certain weak materials, such as unconsolidated sand and mud, may amplify ground shaking at some frequencies owing to their poor elastic response. An example often quoted is the bayfill material around San Francisco Bay. The 1906 San Francisco and 1989 Loma Prieta earthquakes, both of which shook bayfill, provide striking examples of the dangerous behavior of such material. Ground shaking in San Francisco as a result of the 1906 quake was strongest in areas of bayfill (Bay mud Fig. 2.3), as in the Marina district. The only areas shaken more strongly than the Marina district were areas adjacent to the ground-breakage sections of the San Andreas fault. In between the Marina district and the fault, ground shaking was actually less on non-bayfill materials. So the intensity–distance relationship must be qualified by local geological conditions.

The lesson of the importance of material type to ground shaking intensity was repeated during the 1989 Loma Prieta earthquake. Although the epicenter was near Santa Cruz, 60 miles southeast of San Francisco, significant damage occurred in the San Francisco Bay area on structures located on bayfill. Amplification of ground shaking on bayfill contributed to the collapse of the Interstate 880 freeway in Oakland, and to the homes in the Marina district.

Variation in intensity of ground shaking will also occur as a result of the direction from the fault source to the site of ground shaking. If a site is close to the direction of

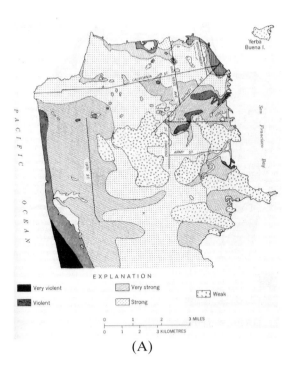

Figure 2.3   Ground-shaking intensity map reconstructed for the 1906 San Francisco earthquake (A); map of the rocks and sediments of the same area (B).

(A)

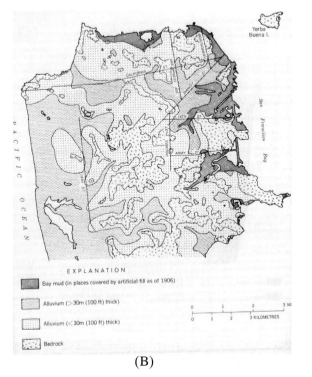

(B)

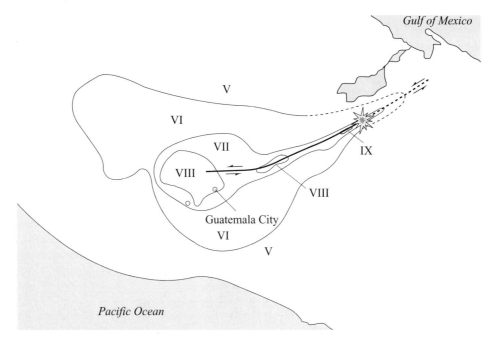

Figure 2.4   Ground-shaking intensity map for the 1976 Guatemalan earthquake. Note that the trend of the fault line with arrows is symmetrical with and approximately parallel to the isoseismals.

rupture propagation, then intensity will be greater. Figure. 2.4 is a map of intensity contours for the Guatemalan earthquake of 1976. The isoseismal lines trend roughly parallel to the surface trend of the fault source. This is a direct result of much of the earthquake energy being directed along the fault trace. Thus, even at great distances from the epicenter, such as at Guatemala City, the intensity of ground shaking is high.

An important type of data used in intensity scales to help define levels of intensity is the effect of ground shaking on structures and objects. Where damage is assessed to determine intensity, care must be taken to evaluate construction type and condition of the structure. Poorly built structures collapse more easily at lower levels of ground shaking, so it is not enough to say the structure collapsed, but how well built it was. This has been a recognized problem in developing useful intensity scales, and it is one reason such scales have been modified with time. The scale used at present in the United States and Western Europe was first developed in Italy by Giuseppe Mercalli near the end of the nineteenth century. It has since been modified or updated owing to changes in construction practice and modes of transportation in the Western world. Today we use the Modified Mercalli scale (Table 2.1). This scale has 12 levels of intensity, which are designated by Roman numerals ranging from the lowest (I) to the highest (XII). The use of Roman numerals makes it easy to distinguish from instrumental magnitude scales, which use arabic numerals, such as the Richter scale.

Data from seismic instruments can be used to locate and estimate the size of earthquakes with much greater precision and accuracy than intensity data. Nevertheless, in-

Table 2.1

THE MODIFIED MERCALLI SCALE OF INTENSITY OF GROUND SHAKING

I. Not felt except by a very few under especially favorable circumstances.

II. Felt only by a few persons at rest, especially on upper floors of buildings. Delicately suspended objects may swing.

III. Felt quite noticeably indoors, especially on upper floors of buildings, but many people do not recognize it as an earthquake. Standing motor cars may rock slightly. Vibration like passing of truck. Duration estimated.

IV. During the day felt indoors by many, outdoors by few. At night, some awakened. Dishes, windows, doors disturbed; walls made cracking sound. Sensation like heavy truck striking building. Standing motor cars rocked noticeably.

V. Felt by nearly everyone; many awakened. Some dishes, windows, etc., broken; a few instances of cracked plaster; unstable objects overturned. Disturbance of trees, poles, and other tall objects sometimes noticed. Pendulum clocks may stop.

VI. Felt by all; many frightened and run outdoors. Some heavy furniture moved; a few instances of fallen plaster or damaged chimneys. Damage slight.

VII. Everybody runs outdoors. Damage negligible in buildings of good design and construction; slight to moderate in well-built ordinary structures; considerable in poorly built or badly designed structures; some chimneys broken. Noticed by persons driving motor cars.

VIII. Damage slight in specially designed structures; considerable in ordinary substantial buildings with partial collapse; great in poorly built structures. Panel walls thrown out of frame structures. Fall of chimneys, factory stacks, columns, monuments, walls. Heavy furniture overturned. Sand and mud ejected in small amounts. Changes in well water. Disturbs persons driving motor cars.

IX. Damage considerable in specially designed structures; well designed frame structures thrown out of plumb; great in substantial buildings, with partial collapse. Buildings shifted off foundations. Ground cracked conspicuously. Underground pipes broken.

X. Some well-built wooden structures destroyed; most masonry and frame structures destroyed with foundations; ground badly cracked. Rails bent. Landslides considerable from river banks and steep slopes. Shifted sand and mud. Water splashed (slopped) over banks.

XI. Few if any (masonry) structures remain standing. Bridges destroyed. Broad fissures in ground. Underground pipe lines completely out of service. Earth slumps and land slips in soft ground. Rails bent greatly.

XII. Damage total. Waves seen on ground surfaces. Lines of sight and level distorted. Objects thrown upward into the air.

tensity data still have usefulness today. Data can be used as a supplement to instrumental data, where the latter information is sparse. This is particularly true for areas where earthquake recording stations are few, and for smaller tremors that are not widely recorded.

Intensity data are also useful in studying earthquakes that occurred before the advent of seismic instruments. The report on the 1680 Malaga, Spain, earthquake is an example of the amount of information that can be obtained by a study of intensity data. The intensity map obtained from the observer reports, combined with a fault map, supplied sufficient data to estimate location (Fig. 2.5), depth (50 km), and magnitude range (M 6.4–7.1).

For many areas in the world, intensity data offer the opportunity to increase the amount of information available in estimating earthquake hazards. In areas where damaging earthquakes occur less frequently, the instrumental data that can be applied to study earthquakes are often insufficient to reach conclusions about earthquake threats.

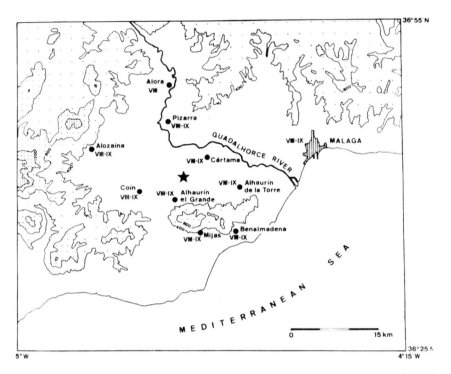

Figure 2.5 Intensity map of the 1680 Malaga, Spain, earthquake. The star shows the estimated location of the epicenter. Roman numerals indicate ground-shaking intensity level.

The intensity data from the pre-instrumental period can provide valuable additional information in the occurrence of large and potentially damaging events.

DEVELOPMENT OF EARLY MECHANICAL SEISMOGRAPHS

The development of instruments to measure and record the motion of the ground during a distant earthquake was a truly great intellectual achievement. Many attempts had been made to devise earthquake instruments over a period of two thousand years before development of true seismographs.

As the term is used today, a *seismograph* consists of three basic components: a seismometer, which responds to ground motion; a chronograph or timing system; and a recording device. The greatest challenge in the development of a seismograph was in the design of a seismometer. This was no easy task, for the goal was an instrument to sense ground motion and therefore not move with the ground.

An object that leaves no permanent record of ground motion gives us no information about an earthquake. For example, a large tree may move as the ground does in several directions during an earthquake, but when the ground comes to rest, because the tree remains firmly attached to the ground, it may show no evidence it has been disturbed. On the other hand, objects not fastened to the ground sometimes do leave a record of ground motion. A heavy chest of drawers in the great Alaska earthquake of

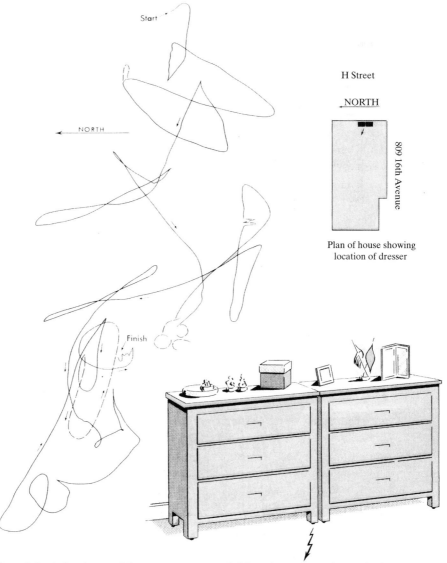

Figure 2.6   A sketch map of the movement pattern left by a dresser leg as the result of ground motion during the 1964 Good Friday earthquake in Alaska.

1964 left a track on the floor etched by its leg (Fig. 2.6). This track is a crude indication of direction of ground movement, chronologically arranged. But it is far from perfect because a certain level of acceleration of the ground is required to overcome friction enough to allow relative movement of the dresser and the ground beneath it. So the dresser is far from being a useful seismograph (Box 2.2.)

The first recognized earthquake instrument was developed in China by Chang Heng in about A.D. 132. China is a country with a very long history of the most devastating earthquakes in the world, so it was perhaps natural for the Chinese to develop the

---

BOX 2.2

Friction, Fluids, and Motion

Friction is the resistance encountered when the surface of one body moves upon or across that of another body. More precisely, the friction between bodies that move is called *kinetic friction*. If the resistance between two bodies prevents motion it is called *static friction*. Friction will vary as the amount of force pressing two surfaces together varies, or as the roughness of two surfaces varies. Fluids also affect the amount of force necessary to cause frictional sliding, for the fluid along a surface under pressure decreases the force acting directly across the surface. Fault slip is easier in the presence of fluids.

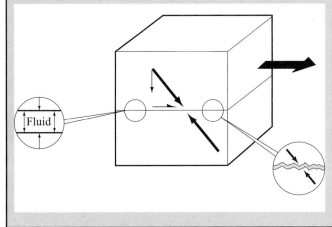

The ability of a liquid to reduce force between two surfaces is due to the relatively incomprehensible nature of fluids. This fact is applied in hydraulic systems to move heavy objects, such as the rudders of jet airliners.

---

first earthquake device. The instrument developed by Chang Heng was not a seismograph by modern definition. A seismograph is a system containing components that make a physical recording of ground motion. Chang Heng's instrument was a *seismoscope*; it responded to ground motion but produced no recording of the movement. His seismoscope consisted of an urn with heads of dragons arranged around the upper edge (Fig. 2.7). Each dragon's mouth contained a ball. When the ground moved, balls located favorably to the direction of ground movement would drop out of the dragon's mouth and fall into the mouth of a frog below. By noting the specific balls dislodged, it was thought to be an indication of the direction to the earthquake. This suggests that Chang Heng had some idea about the spread of earthquake energy along specific paths. However that may be, Chang Heng's device appears to have fulfilled its purpose, for at least one earthquake was detected by this device.

Perhaps early observations that pendulum clocks were often stopped by earthquake ground movements caused some to think of pendulums as earthquake sensors. A pendulum has a number of desirable characteristics as a potential seismometer. First, as already noted it responds to ground motion. What we want is a record of ground motion, reproduced as faithfully as possible. Just as a pendulum may be stopped by ground motion, it can also be set in motion, much as pushing a playground swing, which is a sim-

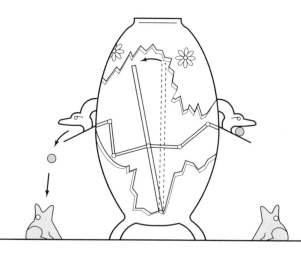

Figure 2.7   Chang Heng's seismoscope as reconstructed from historical documents. This device was reputed not only to have detected a distant unfelt earthquake, but also to have estimated direction to the epicenter.

ple pendulum. Pendulums have the additional advantage that they are motion sensing/responding systems with very little friction involved. The only point of connection with the ground is through the pivot at the top of the pendulum rod (Fig. 2.8). This point can be made virtually friction free, as must be the case for most pendulum clocks. In fact, if this is so, and the pendulum mass is made heavy enough, when the pendulum support above the pivot point is moved, the pendulum may remain virtually motionless, or at least lag greatly behind the motion of the support. Only one simple step remains: to attach a pen or stylus to the pendulum mass so that it lightly contacts a piece of paper to record the motion of the ground beneath the pendulum. Two design principles make this a workable seismometer. First, the pendulum is largely isolated from the ground movement by its suspension design. Second, the mass of the pendulum has inertia, and tends to remain at rest. These two basic principles were utilized in the design of early seismometers.

Two negative characteristics of a pendulum system used as an earthquake sensor are *period* and *resonance*. An ordinary clock pendulum does not make a very practical

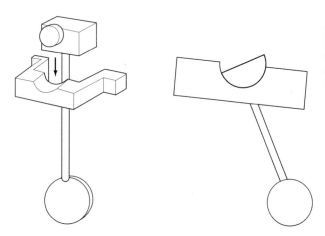

Figure 2.8   Sketch diagram of pendulum attachment points. If properly designed, this point of attachment can be virtually friction free when the pendulum is in motion.

BOX 2.3

Galileo and the Pendulum

Galileo's (1564–1642) contribution to the understanding of the acceleration due to gravity apparently grew out of observations of pendulums. A story often told is that as a young man he attended Mass at a cathedral in Pisa and watched a candelabrum swing to and fro (oscil-

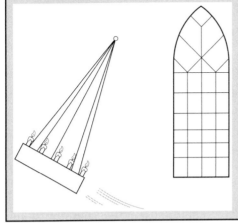

late) after it was lighted and noted that while the swings became shorter, the time of the swing remained constant.

Later Galileo checked his observation by tying stones on a string. He found that with strings of different length, the result was different times of oscillation (free periods), but for a given string length, even with different weights of stones, the pendulum had the same period. Thus, Galileo had discovered that the period of oscillation of a pendulum is directly related to pendulum length.

---

seismometer, because the pendulum is not sensitive to the complete range of ground motions generated by earthquakes. Oscillating systems, such as pendulums, have a characteristic time for one complete cycle or oscillation, known as the *free period* (Box 2.3). In the case of a pendulum, the free period is the length of time in seconds for one complete swing with the rod returning to the starting position (Fig. 2.9). As we shall see, the period of an ordinary clock pendulum does not match the range of periods in earthquake ground motion.

A pendulum will oscillate freely when the period of ground movement is close to the period of the pendulum. The free period of a pendulum is directly related to pendulum length by the formula:

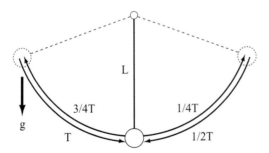

Figure 2.9   The pendulum system, as represented in a clock. The free period (T), a characteristic of any given pendulum system, is controlled by the pendulum length (L). Gravity (g) is generally taken to be constant.

$$T = 2\pi\sqrt{L/g} \qquad \text{where} \qquad T\text{-free period of the pendulum in seconds}$$

L-pendulum length

g-acceleration of gravity (9.8 m/sec²)

We can analyze this equation for the two quantities we are interested in, period (T), and pendulum length (L). If pendulum length increases, then the right side of the equation gets larger; therefore, to retain equality the left side, the period, must also increase. As pendulum length increases, the system responds to longer period ground movement. A grandfather clock typically has a pendulum length of 1 meter (about 3 feet) so that its period is 2 seconds. Large earthquakes can produce ground movement with a cycle of motion (period) of 20 seconds. How long would our pendulum earthquake sensor have to be to match this period? The equation can be solved for length (L), putting in 20 seconds for T:

$$L = T^2g \,/\, 4\pi^2;\; L = (200\ \text{s})(9.8\ \text{m/sec}^2)/39.48 = 99\ \text{meters or } 326\ \text{feet!}$$

Although long pendulums were used by some Italian scientists in the early days of instrumental seismology, a pendulum 326 feet long is not very practical. Fortunately, a new design was devised to solve this problem, the horizontal pendulum. The basic principle is similar to a swinging gate in a fence (Fig. 2.10). If the gate post is tilted off vertical, the gate describes an arc that has not only a horizontal component of movement, but also a vertical one, lifting as it approaches each end of its arclike swing. Thus, we have for all practical purposes a horizontal suspension pendulum. The vertical component of movement is important because gravity, as seen in the pendulum equation, acts against the upward movement of the pendulum in its arclike swing. This arc has a very slight curvature, which suggests that it is equivalent to a vertical suspension pendulum with a very long rod length, and thus a very long period. As very long periods of a ver-

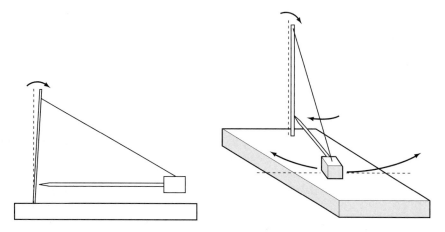

Figure 2.10   The horizontal pendulum operates on the same principle as the swinging gate, which is tilted from the vertical direction. Motion of the mass describes an arc equivalent to that of a very long vertical pendulum.

tical suspension pendulum can be achieved with the horizontal suspension design, we have a very practical solution to our pendulum length problem. The result is an earthquake sensor or seismometer that can respond to long period horizontal ground motion produced by large earthquakes.

The other problem with a pendulum system is the characteristic of *resonance*. Consider again the playground swing as a pendulum. When you push someone on a swing in step—that is, matching the timing or period of your push to the period of the swing—the result is that the swing goes progressively higher, with a greater arc and wilder swings. This is resonance, and not a desirable characteristic in a pendulum system or a seismometer. For a seismometer, the wildly swinging pendulum means that you are only recording the resonant vibration of the system itself, rather than the true motion of the ground. To eliminate this resonance, it is possible to control the pendulum swings, a process called *damping*. Early seismometers had a paddle fastened to the pendulum, which would be immersed in a container of a viscous fluid. As the pendulum moved, the drag of the paddle through the fluid would mechanically "damp" the system, eliminating wild swinging of the pendulum.

The application of the pendulum system as the basis for seismometer design was not easily achieved. It was not until the technological and scientific advances of the eighteenth and nineteenth centuries that the means became available to produce the first workable seismometers, and the first crude seismograph systems.

Andrea Bina in 1751 proposed a common or clock-type pendulum be suspended over a tray of sand, so that the pendulum bob could trace a record of earthquake ground motion in the sand. Although it seems probable that this device was built, it is not known if it ever recorded an earthquake.

The first well-documented pendulum seismometer must be credited to a British scientist, James Forbes. The basis of his seismometer was an inverted pendulum. Tall buildings are in fact inverted pendulums, fixed at their base (see Box 2.4). The Forbes seismometer was installed at Comrie, England, in 1840 to help register a series of earthquakes occurring in the region at the time. It was not very successful. Of 27 earthquakes felt at Comrie in early 1841, only two were detected by pendulum seismometers.

Scientists in Italy were also active in seismic instrument design in the 1800s. Luigi Palmieri spent his scientific career in Italy where he developed a deep interest in both volcanoes and earthquakes. He designed an electrical seismic system that was unique for its time (Fig. 2.11). This remarkable system possessed enough sensitivity to detect earthquakes imperceptible to human beings. The heart of the system consisted of two components, more correctly termed seismoscopes than seismometers. The sensors were able to detect vertical and horizontal ground movement, the initial time of ground movement, its duration, and, like Chang Heng's seismoscope, the approximate direction. However, no permanent record of ground motion could be produced. Because of its success in detecting earthquakes so slight that they could not be felt, the Palmieri seismoscope was widely adopted. It remained the premier seismic instrument until it was replaced by improved pendulum seismometers and seismograph systems near the end of the nineteenth century.

Filippo Cecchi in Italy in 1875 developed what was probably the first workable seismograph system that had as an element a pendulum seismometer. However, no

BOX 2.4

The Skyscraper as a Pendulum

Seismometers are not the only inverted pendulums. Multistory buildings can also act as pendulums, fixed at the ground surface. Because pendulum length determines the free period of a pendulum system, buildings of different heights will have different periods of oscillation.

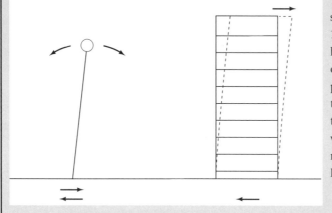

The most intense waves to shake the ground during the 1985 Mexico City earthquake had periods of about 2 seconds. This matched the free period of buildings about 10 to 14 stories high. Some of the buildings were unable to withstand the resulting resonant oscillations and collapsed.

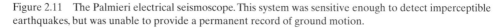

Figure 2.11   The Palmieri electrical seismoscope. This system was sensitive enough to detect imperceptible earthquakes, but was unable to provide a permanent record of ground motion.

record of an earthquake having been recorded by it before 1880 exists. It was apparently less sensitive than those developed after 1875. The first truly successful pendulum seismometers were designed and built in the 1880s by a group of British scientists at the University of Tokyo. Japan in the mid- to late 1800s was a country fully aware of the fact that it was far behind the West in technology. Accordingly, Japan imported scientists to bring the Japanese university science and engineering programs up to date.

British scientists at the University of Tokyo were talented, and within 12 years of their arrival they designed seismometers that would allow the detection of large earthquakes occurring anywhere in the world. The first achievement was the rediscovery and successful application of the horizontal pendulum design by James Ewing in 1880. The horizontal pendulum had been invented by A. Gerard in 1851, but he did not follow through on the design (Box 2.5).

The horizontal pendulum could only respond to horizontal motion of the ground, and so the three-dimensional picture of ground motion was incomplete. The next advance

---

BOX 2.5

Dispute Over Seismometer Design

Paying attention to the men behind the ideas, it is interesting to note that there was some disagreement as to who deserved credit for what among the early inventors of the Tokyo University group of scientists. This resulted in a running debate, at times heated, in the journal *Nature* (1886–1887):

> Dec. 9: Thomas Gray: "For example, he [James Ewing] says, or leads one to infer, that he introduced horizontal pendulums in seismology; not that is not the case.... What Prof. Ewing did introduce was a *particular form* of horizontal pendulum ... but he does not say that the most difficult of the three—namely the vertical components—is written by an instrument which I introduced...."

> Dec. 23: James Ewing: "What I do claim in this matter is that I succeeded in constructing the earliest successful seismograph capable of making absolute measurements of the horizontal motion ...

> "I am not using his [Gray's vertical] instrument.... I have tried to do justice to Mr. Gray's priority in the solution of this problem of vertical astatic suspension; but I prefer, and use my own later solution."

> April 14: John Milne: "By improving the bracket seismograph, Prof. Ewing made a considerable advance in seismometry...;it is hardly fair that his fellow workers, especially Mr. Gray, the most prolific of earthquake inventors, should be passed by unnoticed, and have their work practically appropriated."

> Prof. W. S. Chaplin (Harvard): "I do not remember that in the discussions on your machine [Ewing] Mr. Gray ever claimed to have invented a similar machine, and I am surprised to know that he makes that claim now. On this and other points it appears to me that Messrs. Gray and Milne have not treated your inventions and investigations with fairness."

It should be noted that, today as then, scientific disputes are an important part of the scientific literature in journals. They allow scientific workers to respond to criticism of the work and the scientific community to judge the merits of the debate.

was a completly new design by Thomas Gray. This was the invention of the vertical seismometer in 1881. The approach was to allow the seismometer mass to move vertically by attaching a spring to the mass. The spring replaced the pendulum pivot and decoupled the mass from the ground. This design allowed relative motion of the mass and the support in an oscillating fashion.

The combined use of horizontal and vertical seismometers has resulted in a complete picture of ground motion directions. Standard practice in earthquake observatories has been to use two horizontal seismometers, one oriented north-south and the other east-west, as well as a vertical seismometer.

Early seismograph systems consisted of a seismometer, a clock and timing system, and a recorder. The clock provides absolute time. A number of different ways have been devised to put time marks on a seismic record. Early crude mechanical approaches involved a timed physical deflection of the pen on the record. Modern timing systems consist of crystal-controlled clocks calibrated to Universal Coordinated Time, which in turn are part of an electrical circuit. The electrical circuit will generate a current at the beginning of each time interval (usually the hour and each minute) and this current pulse is transformed into a pen deflection with a device consisting of a coil and a magnet, termed a *galvanometer*. As will be seen, time marks indicating absolute time are critical for calculating earthquake locations, wave velocities, and the time of occurrence of earthquakes.

Different methods have been used to record earthquake ground motion, but the most common has been the recording drum. A rotating cylindrical drum forms the basis of many seismograph recording systems. A motor attached to the shaft that runs through the center of the drum will cause the drum to rotate under the recording pen. Such a motion will leave a line on the paper or seismogram that is wrapped around the drum. A complete rotation of the drum will cause the pen to begin to write over the existing line again. To avoid this, a second motor attached to the pen mount will cause the pen to move parallel to the long axis of the drum, resulting in a spiral line being written on the seismogram (Fig. 2.12). After the paper seismogram is removed from the drum, it can be read like the page of a book. The earliest recording line is on the top left of the seismogram. Time increases to the right on the record, with the last recording located on the last line at the bottom right of the seismogram (Fig. 2.13).

Early seismometers and seismograph systems were termed mechanical because they consisted of clever combinations of springs, levers, rods, and masses. Their primary drawback was that they were not very sensitive and could not record distant earth-

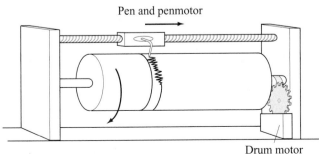

Pen and penmotor

Drum motor

Figure 2.12   The recording mechanism of a seismograph system consists of a rotating drum that moves laterally, or *translates*, as well. The overall motion results in a sequence of lines on the seismogram arranged like the lines on a page in a book.

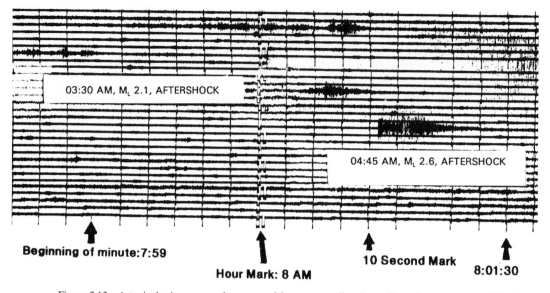

03:30 AM, M_L 2.1, AFTERSHOCK

04:45 AM, M_L 2.6, AFTERSHOCK

Beginning of minute: 7:59

Hour Mark: 8 AM

10 Second Mark

8:01:30

Figure 2.13 A typical seismogram after removal from a recording drum. The seismogram is read like the page of a book: left to right along each line, with time increasing to the right.

quakes well at all. Some magnification of ground motion could be achieved by means of lengthening rods, i.e., increasing pendulum length, or adding rods to magnify the arc of a pen swing. By 1892, John Milne, also at the University of Tokyo, succeeded in manipulating the basic designs to the point where the mechanical pendulum seismometers became more sensitive than the Palmieri seismoscope, and began to supplant it. The Milne seismometer was more successful in detecting distant earthquakes.

Despite improvements, the most successful mechanical seismograph system designs were never able to magnify ground motion more than about 2800 times. This may sound like a lot, but a distant earthquake, or *teleseism*, may not move the ground as much as one hundred millionth of a centimeter. So a magnification of up to 2800 times would not allow detection by the early mechanical systems for many worldwide earthquakes occurring each year. A greater sensitivity was necessary for adequate detection. Such sensitivity would not come from mechanical seismograph systems, but from an entirely new kind of system, the electromagnetic seismograph.

The electromagnetic seismograph was a tremendous advance because the resulting instrument could magnify ground motion hundreds of thousands of times! This invention was the result of work by a Russian aristocrat, Prince B.B. Galitzin, who was a physicist turned seismologist. Prince Galitizin was able to apply the principles of electricity and magnetism to basic seismometer design, so that by 1906 he had developed a working electromagnetic seismograph.

The core of the electromagnetic seismograph concept is that when there is relative motion between a copper wire and a magnetic field, the motion generates a current in the wire. The amount of current in the wire is directly proportional to the amount of motion; the greater the motion, the larger the current. The simple design application of this concept was to wrap a copper wire around the seismometer mass (Fig. 2.14). The rel-

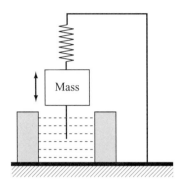

Figure 2.14   The basic principle of the electromagnetic seismograph. Movement of the copper wire attached to the mass, through a magnetic field (dashed lines), generates a current proportional to ground motion.

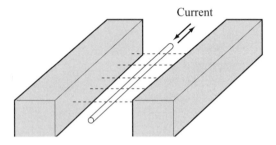

ative motion between a magnet attached to the ground and the copper wire on the mass induces a current. This current can be amplified many times in an electronic amplifier and then turned back into movement of a recording pen by another copper wire/magnet combination (galvanometer) at the recording drum end of the system (Fig. 2.15).

Application of the electromagnetic principle remains the basis for most modern seismograph systems. An important addition to this is the development of digital recording in recent years. Digital recording is achieved by converting voltages into discrete numbers that represent motion by periodically sampling and recording the ground motion as numbers (digits). This can be done automatically and has the advantage that the data are in a much more useful format (Fig. 2.16). Digital data sets representing earthquake ground motion can be used to produce more familiar-looking paper seismogram records directly, or the numbers can be manipulated by computer to select, for example, only certain wave periods, or time intervals, for analysis purposes.

SEISMOGRAPHS AND EARTHQUAKE WAVES

Soon after seismograph systems began recording earthquakes, it became apparent that earthquake waves were complex (Box 2.6). Waves from an earthquake close to a recording station were not only generally stronger but they had an abundance of high-frequency waves (Box 2.7). The opposite was true of distant earthquakes (teleseisms), which had lower frequencies and stretched out across a seismogram for many minutes. It wasn't long before seismologists were separating earthquakes recorded at a station into

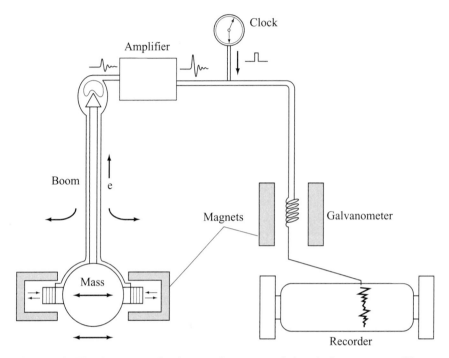

Figure 2.15 The electromagnetic seismograph system consisting of seismometer, amplifier, clock, galvanometer, and recorder.

distance categories based on wave characteristics of the earthquake signature seen on seismograms.

But what of the different earthquake waves represented by the wiggles on a seismogram? The resulting seismograms from earthquakes looked quite complicated. Perhaps the simplest and clearest-looking records came from earthquakes recorded at stations from about 1000 to 9000 kilometers from the epicenter (Fig. 2.17). These records had a clear separation of wave arrivals, usually in a distinct three-part wave package. The first waves to arrive, and thus with the highest velocity, were termed *primae*, Latin for

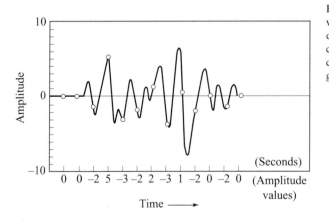

Figure 2.16 Conversion of the wavelike motion of the ground into digits, which can then be stored in a computer. The sampling interval is critical for correct representation of ground motion.

BOX 2.6

The Description of Earthquake Waves

Earthquake waves can be described by using their shape. The wave has amplitude or height and wavelength, which is the distance between two adjacent crests or troughs. The concepts of frequency and cycle, which are related, are also useful in describing earthquake waves.

Let the material disturbed by the passing wave be represented by a mass on a spring. The wave will lift, depress, and lift the mass again, producing a complete cycle of motion. If the motion or cycle repeats itself once in a second, then this is called the *frequency*. The term *hertz* is also often used for frequency, such that 1 hertz (Hz) = 1 cycle/second. The time taken for one cycle in this case would also be 1 second, and is the *period*. Then the two are related by:

$$\text{period} = 1/\text{frequency}$$

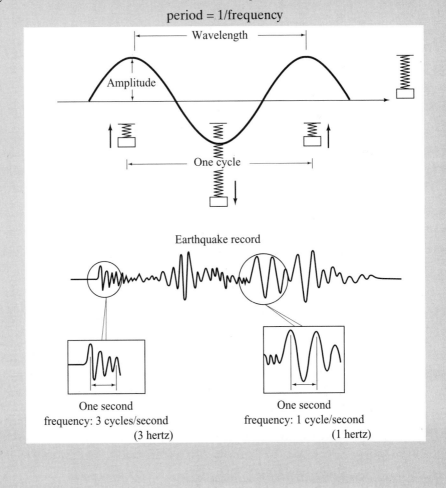

BOX 2.7

The Earth Filter

It has long been recognized that earthquake signatures on seismograms differ with distance from the epicenter. This is because a number of things change as earthquake waves spread out from a focus. These include a relative decrease in amplitude as wave fronts spread out and wave energy is absorbed by rock material, and any increase in length of the signature recorded on the seismogram with increasing distance. The increase of distance leads to two other changes. New waves appear (e.g., Pn) and the frequency and period content of the waves change. Earth acts as a high-frequency filter, absorbing the high-frequency waves more preferentially. Thus, at greater distances we expect to record earthquake waves with generally lower frequencies. The same phenomenon occurs with sound waves in air. Thunder from distant lightning has a low rumble (low frequencies dominate).

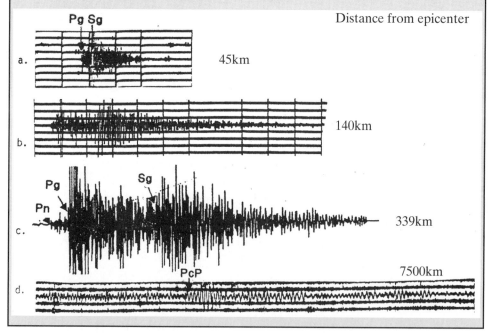

---

primary or first. These were followed a few minutes later by higher-amplitude waves. This second clear wave type was, of course, termed *secundae* or secondary. The abbreviations P-wave (primary) and S-wave (secondary) later became popular for these waves. The third segment of the wave package was the largest and consisted of very high amplitude, long-period waves. Early seismologists called this the *principal section*, and the waves were termed long waves or L-waves.

All of these wave types had in fact been predicted earlier by mathematicians working with elastic-wave theory. As mentioned earlier, it was known that rocks, like other

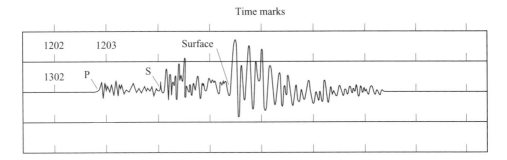

Figure 2.17   A seismogram showing separation of the principal earthquake waves: P, S, and surface waves.

solid materials, can behave elastically. That is, they can respond as a rubber band would if it is stretched. Once the stretching force is released, the rubber band returns to its original shape. The same is true of rock material, although to a lesser extent. If a rock is squeezed in a rock press, it will compress. But while you can easily see a rubber ball compress in your hand, the rock will only change volume (decrease) by a fraction of 1 percent. When a wave, which represents energy and force, travels through rock, the rock deforms elastically by only a small amount. Earth moves, and a seismic record of this event can be produced. Thus, the wiggles on a seismogram can be studied by applying elastic-wave theory.

Elastic-wave theory predicts that in homogeneous solids two types of waves can be transmitted. One is a compressional wave that compresses and stretches the elastic solid in the direction of travel of the wave (Fig. 2.18). The compressional wave is followed by slower waves, which push material from side to side, out of the direction of the wave path. These two wave types were recognized on seismograms by seismologists as the primary compressional waves, and the slower traveling secondary waves.

The P-wave traveling through rock is nothing more or less than a compressional wave. Sound waves are simply compressional waves traveling through the air. This fact explains why observers hear low rumbling sounds associated with earthquakes. When P-waves reach Earth's surface, some of the energy enters the air above the ground and becomes a sound wave. As the wave enters the air, the velocity drops because velocity of seismic waves is governed by the properties of the material it is traveling through. For example, P-waves traveling through rocks near the surface may have an average velocity of about 4.5 kilometers per second. But in the air and near sea level the equivalent sound waves will be traveling at the speed of sound, or about 0.3 kilometers per second. So observation tells us that a decrease in density of material results in a decrease in velocity (Box 2.8).

S-wave velocity is likewise affected by the density and elasticity of material the wave travels through. S-waves distort the medium through which they pass by changing the shape of the material (Fig. 2.18). As the wave passes, elastic rebound restores the shape of the material. The stronger the rock, the more quickly the restoration occurs, and the faster the wave moves. Liquids have no strength, so the wave will not propagate in liquids or gases.

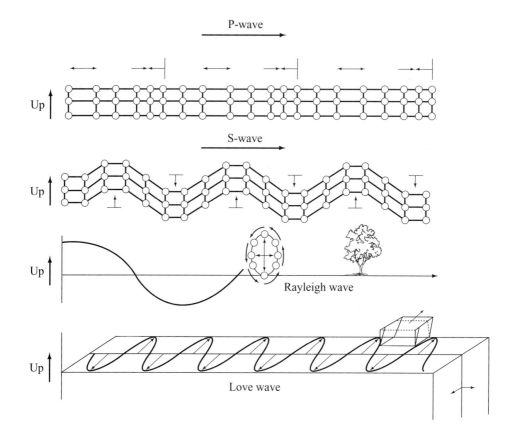

Figure 2.18 Particle motion created by the passage through rocks of the principal wave types.

Waves traveling along the surface of solids had also been predicted before the advent of seismographs. The British mathematicians A. E. H. Love and Lord Rayleigh had predicted the existence of surface waves through application and development of elastic-wave theory. Rayleigh waves can be generated along the bounding surface of a solid material (e.g., Earth's surface) and move rock material in an elliptical pattern with both vertical and horizontal components of motion (Fig. 2.18). The conditions for the existence of the Love wave are more restrictive. Love waves can only be generated where layered rocks exist and where the surface layer is lower-velocity material than that beneath it. Because Love waves generated by earthquakes are observed on seismograms (the principle section), this is proof that the interior of Earth has layers separated by distinct boundaries. The Love wave displaces material like the S-wave, moving rock material from side to side perpendicular to the wave path. However, the Love wave is horizontally polarized—that is, rock material is displaced only parallel to Earth's surface (Fig. 2.18).

BOX 2.8

P- and S-Wave Velocities

The determination of seismic-wave velocities is an important source of data in the study of earthquakes and of Earth. Theoretical and mathematical analysis as well as laboratory studies of elastic materials have resulted in the development of precise relationships between velocity of seismic waves and the physical properties of materials. This allows for the study of the properties of materials that seismic waves pass through deep within Earth.

Movement of particles in rock materials, and the relative resistance of materials to being moved, serves as a test or estimate of physical properties. The arrival of a P-wave wave front compresses and stretches the material in its path, and thus gives an estimate of the material's resistance to a change in volume. This physical property is termed the *bulk modulus* (k). The effect of P-wave energy on moving material particles can similarly be used to estimate the resistance to a change in shape (μ) and density (ρ). The equations that describe the relationship between P-wave and S-wave velocity and physical properties of rocks are:

$$Vp = \sqrt{(k + 4/3\mu)/\rho}, \text{ and } Vs = /\mu\rho$$

The elastic properties of the rock represented by μ, ρ, and k are not independent of one another. As one changes under stress, the others will change as well. Observation shows that a decrease in ρ will result in a simultaneous and more rapid decrease in μ and k, resulting in a decrease in P-wave velocity (Vp).

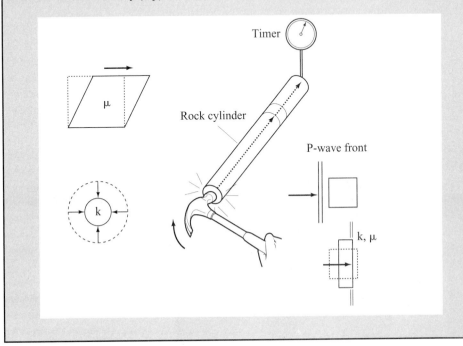

Early seismograms of earthquakes provided the necessary data to prove the existence of all of these elastic wave types within Earth. However, it was not immediately obvious to everyone which wiggle on the seismograms represented a given wave type and was probably the reason for the initial development and use of such terms as *primae, secundae,* and *principal section* for the largest waves on the seismogram.

DEVELOPMENT OF MODERN SEISMIC OBSERVATORIES AND NETWORKS

During the 1880s earthquake observatories were being established in Japan and Europe. The first recordings of earthquakes from a seismograph system in the United States occurred at the University of California (Berkeley) in 1887. The first cooperating network of seismograph stations, spread across the globe, was established by John Milne in the first two decades of the twentieth century. This consisted mostly of stations located in British territories operating Milne seismographs.

The Jesuit Society of the Catholic Church was also very active during the first half of the twentieth century in establishing seismograph stations throughout the world. This was especially true in the United States where long-established stations in Weston, Massachusetts, St. Louis, Missouri, and elsewhere continue to operate today.

The early seismograph stations were usually in close communication and exchanged records, data, and ideas. It was this close cooperation that created the database that would allow the research necessary to understand earthquakes, and the nature of Earth's interior.

Over the years the number of seismograph stations grew slowly. The first real technological leap forward came about 1960 with the establishment of a new worldwide seismograph network, the first one to be established since those developed by Milne and the Jesuits in the early 1900s. Sponsored by the United States, this system was designated the World Wide Standardized Seismograph Network (WWSSN). An advantage of the network was that all stations had the same equipment: seismometers, timing systems, recorders. This allowed easy interchange or combining of data from any and all stations for research purposes. Each WWSSN station was equipped with six seismometers. This enabled more complete recording of ground motion from earthquakes. To obtain a three dimensional record of ground movement at least three seismometers were necessary: vertical, north-south, and east-west (Fig. 2.19). These three components were du-

Figure 2.19   Vault of a typical seismograph station. The vault contains three short-period and three long period seismometers.

plicated at each station to cover a greater bandwidth of frequencies. Three components were long-period with seismometers designed to respond best to earthquake waves ranging from about 5- to about 100-second wave periods. The other three seismometer components at a WWSSN station were short period—that is, the seismometer design is such that the response to ground motion was best in the period range from about 0.05 to 5 seconds (Fig. 2.19).

Significant developments have taken place in the design of seismograph systems since 1970. The most important advances have involved seismometers and recording systems. The standard WWSSN system with its six component seismometers lacks complete recovery and recording of ground-motion information. This is true for two reasons. First, the combination of short- and long-period seismometers available in the 1960s was not adequate to give high-gain recovery of ground motion at all possible periods. The WWSSN systems were in fact designed to filter out a high level in non-earthquake noise in the frequency range of 0.1–1 hertz, i.e., 1- to 10-second wave period (Fig. 2.20). The strategy was to choose peak magnification of ground motion above and below these frequencies, leaving a low-gain gap in the area of the increased noise. This is no longer necessary because of the introduction of digital filtering to eliminate noise.

The second reason that WWSSN stations were limited in recovery of complete ground motion is related to design. Until recently, most seismometers have been designed as inertial sensors. The inertial sensor consists of a mass that tends to remain stationary or inert, and when movement of the ground occurs there is relative displacement between the frame or support and the mass. There are drawbacks to such designs.

One drawback of the inertial seismometer is that for a large earthquake, often a station could not record complete ground movement because the amplitude of ground motion was greater than the pendulum system could respond to. Consequently, the result would be a recording on the seismograms of waveforms with clipped peaks (Fig. 2.21).

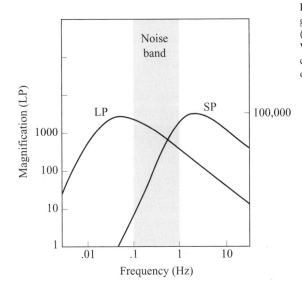

Figure 2.20   Peak magnification of ground motion for the long-period (LP) and short-period (SP) WWSSN seismometers in comparison to the frequency band of noise.

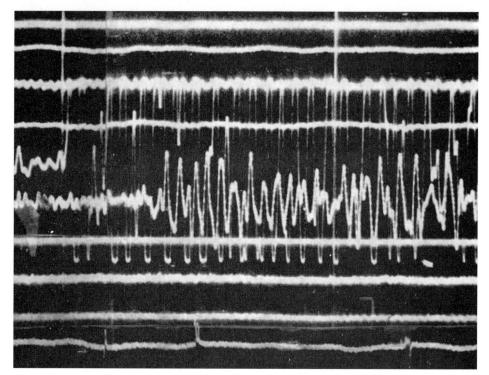

Figure 2.21   An earthquake recording showing clipped (square) wave peaks owing to a lack of adequate dynamic range in the seismograph system.

The new broadband seismograph systems utilize a force feedback design to increase seismometer stability. An electric current proportional to the displacement the mass undergoes is put into an electronic circuit. It is converted to a proportional force that is applied to the mass and opposes its relative motion. Thus, because of this design strategy, broadband force feedback seismometers are very stable under a wide range of forces (dynamic range), are more sensitive, and can respond to a wide range of wave periods (Fig. 2.22).

Equally impressive advances have been made at the other end in modern seismograph systems. For nearly a century the primary recording medium in seismograph systems has been paper. This has included a number of variants such as ink, photographic paper, smoked paper, and chemically treated heat sensitive paper. New methods of data acquisition and storage have been made possible by advances in computer technology. These methods in turn made possible new techniques of analysis of earthquake waves.

A very useful and important technique is the separation of earthquake waves by frequency. This is difficult to do from a paper record, because the waveforms must be processed or digitized by hand. Amplitudes must be sampled and measured from the entire earthquake signature at equal time intervals (e.g., 1 second). The record of the earthquake waves is converted into numbers representing position (amplitude) and time (Fig. 2.16). This can be very tedious and time-consuming especially if one has 20 or 30 station records to convert.

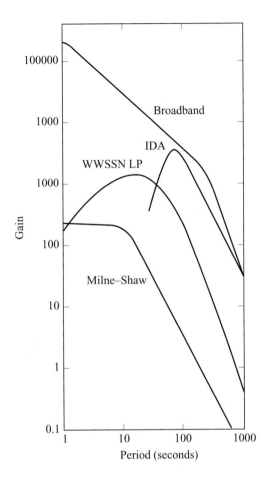

Figure 2.22   A graph showing the range in magnification of ground motion with frequency of different seismograph systems. The modern broadband system is a significant improvement on previous designs.

Consequently, before the advent of sophisticated computers and digital recording, such research techniques were not very frequently applied. With digital recording the data are automatically collected and stored in computers in the appropriate format. The important parameters of a digital recorder are the sampling interval (in seconds), the voltage that is equal to a change in amplitude/unit (Fig. 2.16), and the ratio of the largest to smallest number that can be recorded (dynamic range). The data can then be stored in a computer or on digital media such as magnetic tape, processed by computer programs, and paper records printed out as needed.

DIGITAL NETWORKS AND ARRAYS

Digital recording of ground motion from earthquakes is one of the most significant advances in earthquake instrumentation since 1900. An advantage of digital recording is the ability, once the data are in the computer, to filter out or reject frequencies that represent noise, thus increasing the signal clarity. This may also be done by combining the outputs of geographically separated but clustered seismometers. The noise will be dif-

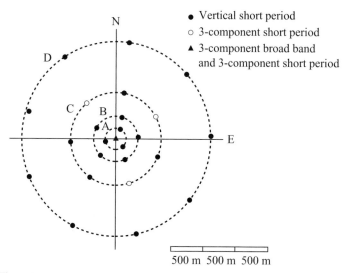

- • Vertical short period
- ○ 3-component short period
- ▲ 3-component broad band and 3-component short period

500 m 500 m 500 m

Figure 2.23 Map view of the geometric configuration of seismic arrays.

ferent at each station, and when the outputs of different seismometers are combined, the noise will cancel out, because it is random and variable.

Seismograph clusters are termed *arrays* and can be used to determine velocity of seismic waves and direction of the source, allowing locations to be determined (Fig. 2.23). Arrays can vary in size. The LASA (Large Aperature Seismic Array) in Montana had seismometers spaced as much as 200 kilometers apart. Such arrays can be used to discriminate between earthquakes and nuclear explosions and to monitor nuclear test ban treaty compliance.

Networks using digital recording and broadband sensors have banded together throughout the world to form the Federation of Digital Broad Band Seismograph Networks (FDSN). The cooperation of network operators within the federation results in increased availability of digital broadband data from earthquakes for research workers worldwide (Fig. 2.24).

SUMMARY

The first data available to study earthquakes were observer reports of ground-shaking intensity and damage to structures. This resulted in the development of intensity scales of ground shaking. Maps of these intensities were used to estimate epicentral locations and earthquake size. Today, the Modified Mercalli intensity scale (MM) is used widely in the United States and Western Europe. Intensity mapping is presently used as a supplement to instrumental earthquake data and in earthquake hazards studies. It is also used to study historical pre-instrumental earthquakes.

Development of modified pendulums to measure ground motion began in the early 1800s and culminated in the first successful instrument to measure and record the

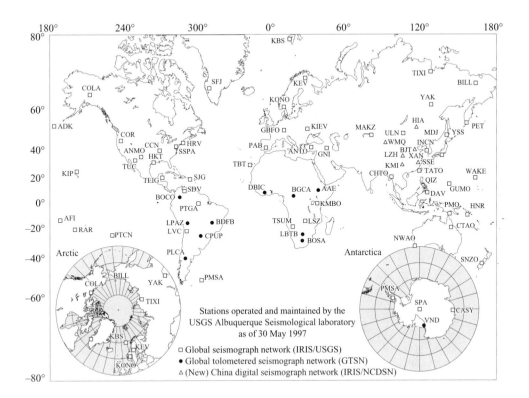

Figure 2.24   Digital worldwide stations operated and maintained by the United States Geological Survey Albuquerque Seismological laboratory.

time history of ground motion in earthquakes in the late 1800s. A group of British scientists in Tokyo demonstrated successful sensor (seismometer) and recorder systems (seismographs) that were subsequently widely adopted.

Analysis of early earthquake records (seismograms) resulted in the identification of several wave types predicted by theoreticians: the primary (P) wave, secondary (S) wave, and the Love and Rayleigh surface waves.

The mechanical seismograph designs of the 1800s were superseded by electromagnetic seismographs in the early 1900s. Electromagnetic seismographs convert pendulum motion into electric voltage. The electric current resulting from the voltage can then be amplified thousands of times to dramatically increase seismograph sensitivity. This basic design application underlies modern seismograph systems.

The state-of-the-art seismograph systems in use throughout the world include broadband seismometers that can respond to a wide range of ground motions of different periods. The amplified outputs of such systems are recorded digitally, which allows for analysis by use of computer software.

Groups of seismographs are termed *networks*. The early Milne network was superseded by the World Wide Standardized Seismograph Network (WWSSN). Establishment of digitally recording broadband stations has led to the development of the Federation of Digital Broad Band Seismograph Networks (FDSN). Networks of seismometers with special geometric distributions of sensors are termed *arrays* and are used for special purposes such as nuclear test detection and research on the structure of Earth's crust.

KEY WORDS

array	Forbes	Lord Rayleigh
Bina	force feedback	Love
broadband	friction	magnification
Cecchi	Galitzin	Milne
Chang Heng	Gray	Mercalli
chronograph	inertia	noise
damping	inertial sensor	Palmieri
digital	intensity	pendulum
elastic waves	isoseismal	recorder
epicenter	Jesuit	resonance
Ewing	L-waves	seismogram
focus	long period	seismometer

C H A P T E R   3

Faults and Earthquakes

AN INTRODUCTION TO FAULTS AND FAULTING

Although ground shaking can occur from many causes (such as volcanic eruptions), clearly the great majority of earthquakes, including the very largest, are related to movement along faults. So in order to understand earthquakes, we must also understand faults and faulting.

A *fault* is a shear fracture. The rocks on one or both sides of the fracture surface will have slipped along it. Such a simple, dry definition cannot begin to express the power implicit in the existence of a fault like the San Andreas in California. A careful study of the movement of such a fault during an earthquake demonstrates the power and energy involved. Consider for a moment the weight of a cube of rock 1 meter (approximately 3 feet) on a side. A typical piece of crustal rock has a density of 2.7 grams per cubic centimeter. There are a million (10^6) little cubes 1 centimeter on a side in a 1-cubic-meter volume of rock. Thus, the 1 cubic meter of crustal rock would weigh 2.7×10^6 grams, or in English units approximately 3 tons! Now imagine the weight of rock moved along the San Andreas fault during the 1906 San Francisco earthquake. Fault slip occurred along 430 kilometers of the fault in less than a minute. If this movement is restricted to a width across the fault of 100 meters (0.1 kilometer), and a depth down to about 10 kilometers, then the volume of rock moved is 430 cubic kilometers (4.3×10^{11} cubic meters). Given 2.7×10^6 grams/cubic meter, the mass of rock that moved would be at least 11.6×10^{17} grams (or approximately 1290 billion tons!). During the time of the earthquake this mass was actually accelerating. This expression of power is easy to see along the San Andreas fault. Offset stream drainages and outcrops of distorted rock in road cuts bear testimony to the incredible forces unleashed along faults like the San Andreas during an earthquake (Fig. 3.1).

NOT ALL FAULTS ARE ALIKE

The San Andreas fault marks the boundary between two gigantic plates of rock of Earth's crust. These plates are in very gradual motion, such that the rocks on both sides of the fault are moving horizontally past one another. This is why the offset of features that cross the fault, such as streams, can be seen so well. Such faults are called *strike slip*. Other examples of strike-slip faults include the great North Anatolian fault in Turkey; the Great Glen fault in Scotland, which traverses Loch Ness; and the Alpine fault in New Zealand (Fig. 3.2).

Figure 3.1   Roadcut showing the tremendous deformation of rocks created by forces generated along the San Andreas fault.

Faults, such as strike-slip faults, are grouped together by their characteristics. These include inclination of the fault surface and the slip direction. Thus, the term *strike slip* is a concise way of describing the San Andreas fault. The slip direction is parallel to the strike of the fault. Strike is simply the compass direction of the fault line on the ground at Earth's surface measured from north. The inclination or dip is the acute angle (less than 90 degrees) between the fault surface and Earth's surface, measured perpendicular to strike (Fig. 3.3). For a strike-slip fault, the inclination or dip is usually nearly vertical (90 degrees).

There are two possible kinds of relative motion on a strike slip fault, termed *right lateral* and *left lateral*. Looking along the strike of the fault, if the left side block has moved toward you, it is left lateral. Similarly, if the right side block has moved relatively toward you, it is right lateral. The Great Glen fault in Figure 3.2 is an example. Past fault motion is indicated by the arrows along the fault as well as by the offset of the rock cut by the fault (stippled pattern). Looking along the fault from either end, it can be seen that the left-hand block or side has moved toward the viewer. Thus, the Great Glen fault is left lateral.

Fault motion is possible in directions other than the strike of a fault, such as in the direction of dip. These faults are called *dip slip*. One block may be literally pushed up and partly over another. This is what happened in the Northridge, California, earth-

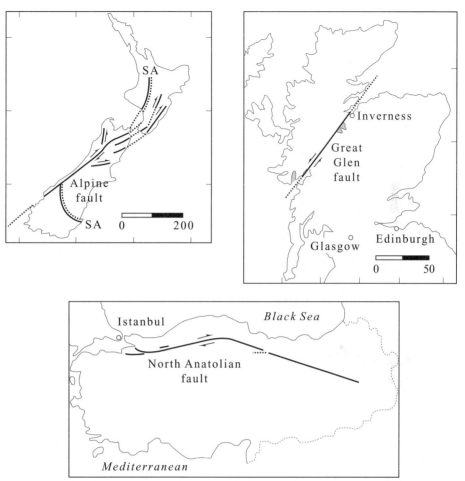

Figure 3.2   Maps showing large strike-slip faults in New Zealand (Alpine fault), Scotland (Great Glen fault), and Turkey (North Anatolian fault).

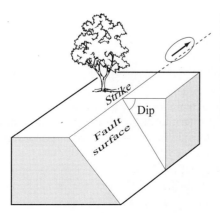

Figure 3.3   The strike of a fault can be measured as the compass direction of its trace at Earth's surface. *Dip* is the angle less than or equal to 90° between the horizontal (Earth's surface) and the fault surface, measured perpendicular to strike.

quake of 1994. Such faults are called *thrusts* or *reverse faults*, depending on how steeply the fault surface is inclined.

Active thrust faults occur in the Los Angeles basin. The 1971 San Fernando earthquake represented energy released by movement on a thrust fault that broke the surface of Earth. Other large thrusts are to be found along the base of mountain fronts such as the Rocky Mountains of the western United States and Canada. The Lewis thrust fault, which is no longer active, lies just to the east of Glacier National Park and stretches from there into southern Canada. At Marais Pass just south of Glacier National Park a cross section or side view of the Lewis thrust is exposed by the deep erosion producing the pass (Fig. 3.4). The fault surface at this location is seen to be nearly flat and is overlain by rocks that can be found to the west in wells, thousands of feet below Earth's surface. Movement along the Lewis thrust has brought this enormous volume of ancient and formerly buried rocks to the surface. The tremendous energy required to perform this feat was supplied by many thousands of large earthquakes over millions of years (Box 3.1).

Thrust-fault movement results in overall shortening of Earth's crust. If crustal rocks are stretched or expand horizontally then another kind of fault will result, called a *normal fault*. To accommodate increased space from expansion, the block above the fault, termed the *hanging wall*, will slip downwards relative to the underlying foot wall block (Fig. 3.5). Such faults can occur at all scales, including slip over millions of years equivalent to several thousand feet of change in elevation. A large group or system of normal faults stretches north-south across Utah. This is known as the Wasatch fault zone and separates the higher Colorado Plateau to the east, from the lower deserts of Utah, which lie to the west. The north-south zone of normal faulting continues across Utah and on into the vicinity of Yellowstone National Park where in 1959 the Hebgen Lake earthquake caused landslides that killed 28 people. Fault surface rupture and slip created scarps as high as 21 feet (Fig. 3.6).

FAULT SURFACES AND PROCESSES

Movement on faults occurs in at least two different ways: by stick-slip or unstable frictional sliding, and by stable sliding or creep. The stick-slip mechanism involves sudden movement on the fault after accumulation of stress. This mechanism and the accompanying elastic rebound are what we associate with earthquakes.

Figure 3.4   Marais Pass, Montana, with a view to the north at the exposure of the fault surface of the inactive Lewis overthrust fault.

BOX 3.1

Force, Stress, and Strain

The application of force to rocks will often lead to permanent deformational features such as faults. Although the total amount of force is important to understand, the intensity of force is also important, because it is a useful way of describing the potential for deformation. Intensity is expressed by force per unit of area of the surface to which the force is applied. Force per unit of area is known as *stress*. An example might be the force applied through a person's arm to his hand spread on a table top. The table top does not deform in part because the force is spread out over the area of the hand. The force becomes more intense when concentrated in the point of a sharp ice pick. In the latter case the intensity, or stress, is greater because the area of application is smaller, and the result is deformation of the table top or permanent strain.

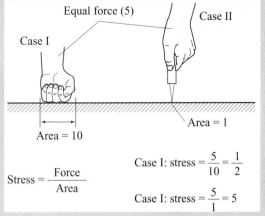

$$\text{Stress} = \frac{\text{Force}}{\text{Area}}$$

Case I: stress $= \dfrac{5}{10} = \dfrac{1}{2}$

Case I: stress $= \dfrac{5}{1} = 5$

The culprit in the storage and rapid release of earthquake energy is friction. Friction resists movement on fault surfaces, allowing strain to build up in the rocks around the fault. But when the level of stress exceeds frictional resistance, fault slip will occur, the rocks rebound elastically, and earthquake waves radiate outward.

Friction along a fault surface can exist for several reasons. A certain amount of friction will always exist on a fault because of the weight of the overlying rocks pressing against it. The more horizontal the fault surface, the greater will be the force from the weight of the overlying rocks. The increased weight will result in increased friction. It was once thought that nearly horizontal thrust faults, like the Lewis thrust, could not have movement on them because so much force would be required to overcome friction and move the rock that the rock would be crushed first.

Water is the answer to the motion-versus-friction paradox. Because water is relatively incompressible, if water is trapped along the fault zone it supports the weight of the rocks and decreases the force across the fault surface, helping to make movement possible at low fault inclinations (Box 3.2). Fault movement on dry surfaces requires higher angles to overcome the friction. Eventually any block on a surface becomes unstable and will slip if the surface inclination is high enough.

Friction can also exist because of the shape of the fault surface. Most fault surfaces are neither completely flat nor smooth. A close examination of fault surfaces will reveal rough spots, or *asperities* (Fig. 3.7). Asperities increase frictional resistance to movement beyond that which exists on a smooth surface by serving as barriers to movement. If movement is to continue, the asperities must be broken to smooth out the fault

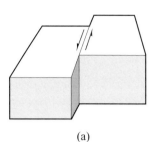

(a)

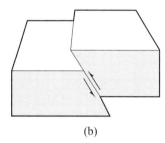

(b)

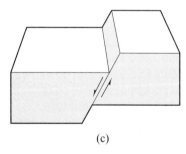

(c)

Figure 3.5   The three principal types of
faults: (A) left-lateral strike-slip; (B)
reverse; (C) normal.

surface. Even during movement in an earthquake, a large enough asperity may change
the rupture pattern or stresses on a fault surface, or even stop the motion altogether, end-
ing the quake.

The rupture front on a fault surface will begin at one point, which is termed the
earthquake focus. Beyond that point the advance of the front may be quite complex, affected in part by asperities on the fault surface. The 1989 Loma Prieta earthquake is a
good example of the complex nature of slip on a fault plane in a large earthquake. As
seen in Figure 3.8 not only the amount of slip, indicated by the length of arrows, but the
direction of slip may vary. The areas of highest slip and greatest stress release probably

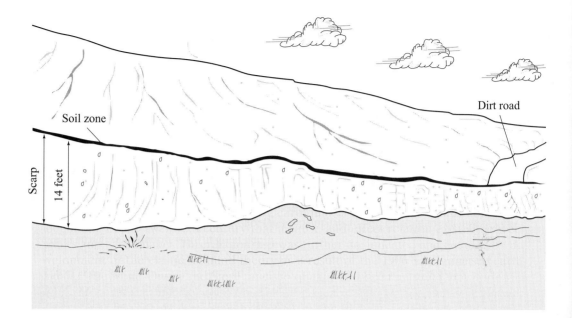

Figure 3.6   Fault scarp created by the 1959 West Yellowstone/Hebgen Lake earthquake.

represent the location of asperities broken during the faulting process. Note from this figure that during this earthquake the rupture did not reach the surface.

The curvature of fault surfaces in three dimensions may also increase friction and create barriers to movement. The 1983 Borah Peak, Idaho, earthquake involved movement on a normal fault surface, the Lost River fault. The release of stress began at a bend in the fault (Fig. 3.9). This bend also served as a barrier, eliminating any movement to the southeast. The identification of bends separating faults into segments is an element in earthquake hazards studies.

Bends that result in a vertical rise of a fault surface may also be present, affecting friction and motion (Fig. 3.10). Vertical bends are called *steps* or *ramps* and can occur on both normal and thrust-fault surfaces. A thrust-fault ramp increases the amount of force needed to overcome friction because the rocks above the fault surface must move upward against the force of gravity.

Bends may also occur along the strike of vertical strike-slip faults. Perhaps the most famous bend in a fault occurs on the San Andreas fault. This is the Big Bend, where the San Andreas turns to the west (Fig. 3.11). Because of the bend, the block containing Los Angeles on the west side, in attempting to move northwards, compresses the rock against the fault, shortening it against the San Andreas. The result is that a number of thrust faults are active under the city of Los Angeles, accommodating this shortening.

The Big Bend is a turn to the left. What would happen on the same fault if the motion of the blocks were reversed, with the Los Angeles block moving southward? The area of the bend would open up, creating a hole that would then collect sediment brought in by streams. This is the situation that has developed in Siberian Russia. The

BOX 3.2

Turning on Earthquakes

The role of water in the earthquake process was emphasized by an unintended natural experiment near Denver, Colorado, in the 1960s. Denver is located in a relatively earthquake-free area, so when the region began to experience tremors in 1962, it created public concern. The earthquakes were located around an Army waste-disposal well. The well crossed a fault surface, and whenever liquid waste under pressure was injected into the ground, earthquakes occurred. The pressurized fluid moving along the fault, which the well cut across, reduced friction and triggered fault slip and earthquakes.

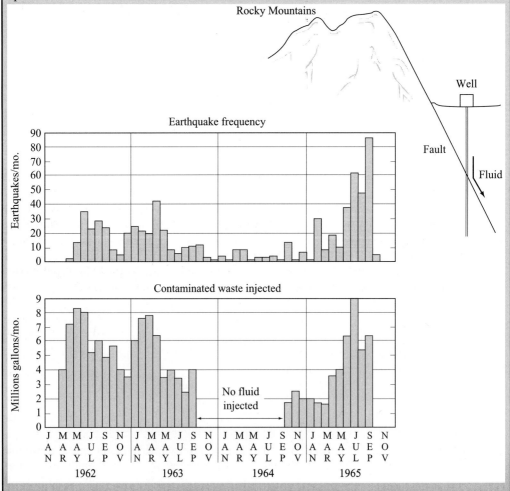

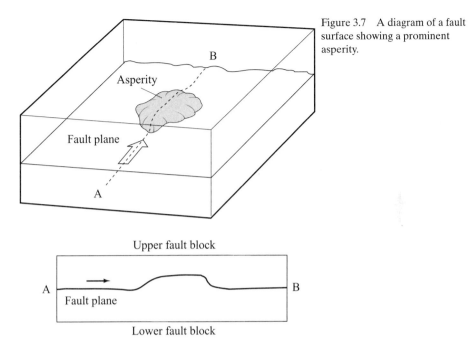

Figure 3.7   A diagram of a fault surface showing a prominent asperity.

hole at the strike-slip bend is occupied by Lake Baikal, the deepest lake in the world (Fig. 3.12).

Available water under pressure along fault zones may lower friction to the point where force cannot be built up, because the fault will move slowly and continuously. This is termed *creep* or *stable sliding*. In such a situation stress is being released continuously without being built up to a level where an earthquake would result. This is in contrast to friction-driven buildup of stress or stick-slip movement. For some faults both kinds of movement and stress release occur.

The primary example in the United States is the San Andreas fault, which stretches across California for a distance of 1120 kilometers (700 miles). Along the trend of

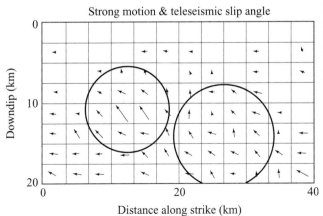

Figure 3.8   Northwest-southeast cross section along the fault-plane model of the San Andreas fault showing slip directions and amounts during the 1989 Loma Prieta earthquake.

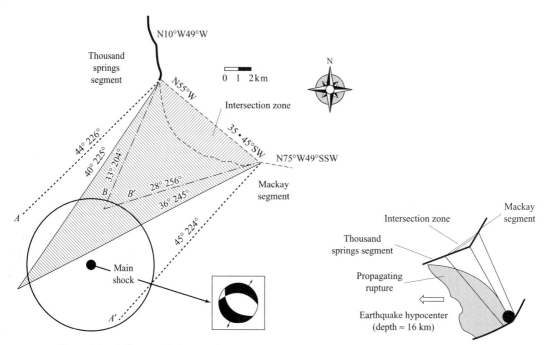

Figure 3.9 A diagram of the overall geometry of the Lost River fault and the position of the earthquake focus for the 1983 Borah Peak earthquake. Also shown is the relationship between the focus and the fault segments.

this fault different kinds of earthquake activity and fault movement occur. Near Hollister, California (Fig. 3.11), stable sliding or creep occurs, while elsewhere stick-slip movement with accompanying earthquakes is more the norm. The movement along the San Andreas is predominantly horizontal (strike-slip) with individual earthquakes such as the 1906 San Francisco tremor contributing as much as 20 feet (6 meters) of slip. The overall effect is to move the land surface west of the fault (e.g., Los Angeles) northward relative to the landscape east of the fault. Cumulative movements over a period of millions of years have totaled 350 miles. The most impressive movements have occurred during the great earthquakes. Historically this would include the 1857 magnitude-8 earthquake near Ft. Tejon, and the 1906 San Francisco earthquake. These events have been well publicized; however, many other damaging earthquakes have occurred along the San Andreas in the last 200-plus years of recorded California history.

FAULT BEHAVIOR AND TIME

Larger earthquakes are difficult if not impossible to predict at present, but along the central San Andreas fault, at Parkfield, California, moderate earthquakes appear to occur with almost predictable frequency. Over the period 1857 to 1966 the San Andreas moved

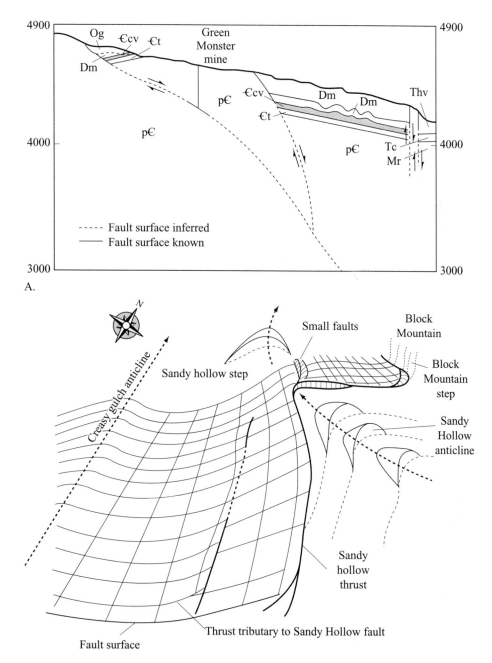

Figure 3.10 Vertical bends termed *steps* or *ramps* in normal (upper) and thrust-fault surfaces (lower). (A) Cross-section view of a step on the Verde normal fault, intersected by the Green Monster mine shaft. (B) Step geometry of the Sandy Hollow thrust fault surface (grid pattern).

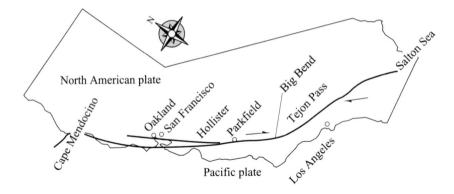

Figure 3.11   The Big Bend on the San Andreas fault. This change in strike of the fault causes forces to build up along this section of the fault, and this no doubt increases earthquake hazards in the vicinity.

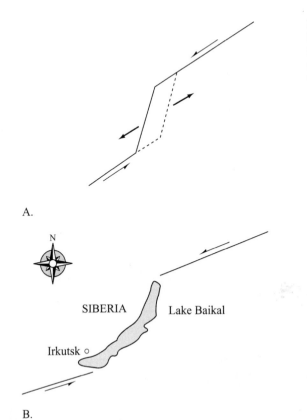

Figure 3.12   A sketch map of Lake Baikal, Russia. The fault bend here results in a basin filled by the deepest freshwater lake in the world.

A.

B.

and generated moderate earthquakes (M 5.5–6.5) on the average of every 22 years: 1857, 1881, 1901, 1922, 1934, 1966. Although there is an unusually long 32-year gap from 1934 to 1966, the other intervals come close to the 22-year average. This is a very regular cycle of earthquake activity and suggests a regular buildup and release of stress on this segment of the San Andreas. This kind of behavior is unusual as can be seen from a study of historic earthquakes elsewhere on the San Andreas.

Much of the northern section of the San Andreas from Cape Mendocino to Hollister (Fig. 3.11) has a relatively high rate of activity, generating a few large damaging earthquakes, as well as more frequent smaller events. None of these seem to occur in regular cycles as at Parkfield. The southern section of the San Andreas from Tejon Pass to the Salton Sea does not show as high a rate of activity as the northern section but still slips, generating earthquakes in sequences in which regular cycles have not been recognized. The large 1857 Ft. Tejon earthquake occurred on this part of the San Andreas, and study of this segment of the fault indicates that future large events will occur.

The central section of the San Andreas near Hollister does not exhibit the classical stick-slip model of stress release. It is relatively quiet with few earthquakes. This is because stress is being released by stable sliding or creep, releasing stress gradually and continuously. Ample evidence of this can be seen in Hollister, where sidewalks, fences, porches, and street curbs are offset, repaired, and offset again, over a period of years owing to the slow, steady movement of Earth's surface (Fig. 3.13).

A classic example of the effects of creep can be found at the Cienega winery, which is actually built over a creeping section of the San Andreas fault. Here slow movement

Figure 3.13   Offset curb in Hollister, California, caused by creep movements on the San Andreas fault.

Figure 3.14   Drainage ditch offset adjacent to Cienega Winery by creep on the San Andreas fault.

of 1.5 centimeters per year on the fault over years has warped the walls of the winery building with dramatic results. Equally impressive is the separation of a concrete-lined drainage ditch outside the structure (Fig. 3.14).

FAULTS AND TOPOGRAPHY

Whether movements are rapid or slow, the net effect for faults like the San Andreas is deformation of Earth's surface. Abundant evidence exists along the length of the San Andreas of fault movement and resulting deformation. Over long periods of time surface topography is dramatically changed. Streams that cross active faults have their channels offset, and the drainage will follow the fault for a distance. If the movement is rapid enough, the stream is beheaded, or cut off from its source. Because the fault zone itself is an area of shattered and weakened rock it is subject to erosion, and is often manifested as a topographic valley, trapping water as lakes or ponds termed *sag ponds* (Fig. 3.15).

Dip-slip faults with dominantly vertical motion create mountain block topography with sharp breaks in slope at their base where the fault is located. Perhaps the most famous and beautifully scenic example of this type of topography is the mountain front of the Grand Teton Range in western Wyoming (Fig. 3.16). The topographic low at the base of the mountain range caused by faulting is utilized as a reservoir created by construction of a dam downstream from this location. The rugged mountainous topography

Figure 3.15   A view of a sag pond located along the trace of the San Andreas fault. Erosion along the fault trace produces topographic lows that trap water.

and presence of continuing small earthquakes strongly suggest that the Teton fault is still active.

Fault scarps are useful as an additional source of data in understanding earthquakes (Box 3.3). The deformation of the ground by offset or slip that can be measured on a fault scarp can be used to model the overall slip on the fault from an individual earthquake. The basic tools for such a model analysis come from the science of *geodesy*, which is concerned with the location or change in location by deformation of points on Earth, both horizontally and vertically. A geodetic model analysis was conducted after the 1989 Loma Prieta earthquake. Vertical changes in elevation of the ground due to subsurface fault slip were measured by comparing positions around the epicenter, which were obtained in geodetic surveying before the tremor, to a survey of positions completed after the quake. These data were used to obtain an average fault model for the

Figure 3.16   View west across Jackson Lake at the fault-controlled front of the Grand Teton Mountains of Wyoming. The flat slopes near the mountain base represent erosional remnants of the fault scarp. This fault is active.

tremor. Some constraints may be placed on the geodetic modeling by calculations obtained from seismograph records of parameters such as total slip and fault area (Fig. 3.17).

In areas where there is a low frequency of damaging earthquakes, little primary data is available from seismographs to study earthquake activity. This is a situation where examination of surface fault scarps, and ancient buried fault scarps, has proven useful. Fault scarps are produced by larger earthquakes (M > 6.5), and it is precisely these events that are of interest with respect to earthquake hazards analysis.

A large earthquake occurred in northern Sonora, Mexico, in 1887. The quake was felt strongly as far north as Phoenix, Arizona. But because it occurred before the installation of operating seismometers in North America, its magnitude was unknown. However, techniques now exist that use fault displacement and fault length to estimate earthquake magnitude. A surface fault scarp was formed in the 1887 earthquake that was 80 kilometers long. This scarp can be seen today, more than a century later (Fig. 3.18).

Equations have been developed that relate earthquake magnitude to fault length from the study of instrumentally recorded earthquakes and the corresponding fault rupture. Thus, for pre-instrumental earthquakes, as well as prehistoric earthquakes, fault length can be used to estimate instrumental magnitude. For the Sonoran earthquake, a scarp length of 80 kilometers indicates a magnitude of 7.4. Equally interesting, a study of the area of the 1887 earthquake has revealed a second scarp near the original rupture, indicating yet another large earthquake. This information is invaluable in earthquake hazards studies because age dating of scarps gives an idea of how long the intervals are between large damaging tremors. Prehistoric earthquakes are often called *fossil* earthquakes or *paleoearthquakes* because features like fault scarps are preserved as a record of their occurrence. The general study of pre-instrumental and prehistoric earthquakes from evidence left in the rocks and soils is termed *paleoseismology* (see Chapter 10).

HIDDEN FAULTS

Earthquake hazard would be easier to assess if all earthquake faults were exposed at the surface, like the San Andreas or the Teton. Unfortunately, this is not the case. Small earthquakes of magnitude less than 5 almost never rupture Earth's surface to produce

BOX 3.3

Fault Scarps While You Wait

An important advance in the study of earthquakes was the realization that there was a link between the occurrence of an earthquake and the formation of a fault scarp. However, formation of a fault scarp during an earthquake is an event that takes place so quickly that actual observation of the formation of a fault scarp is a rare event indeed. This did occur, however, during the 1983 Borah Peak, Idaho, earthquake. Although the area where the earthquake occurred was sparsely populated, fortuitously there were a number of people out in the fields and hills owing to the opening of hunting season. Several observer reports were printed in the *Seismological Society of America Bulletin*, and are worth reproducing here:

A most remarkable report is that of two hunters in a Bronco who were driving down a jeep trail. When the fault scarp formed during the earthquake, their vehicle was 45 feet away from it (⊕ on map):

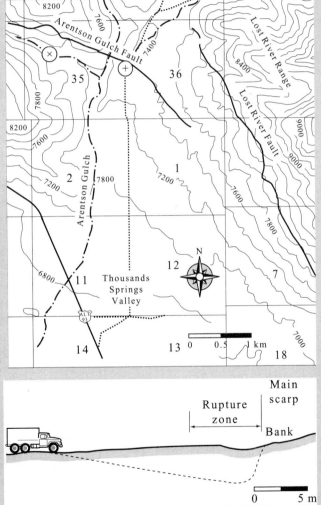

Hendriksen: "All of a sudden I did feel light-headed and I just lost my equilibrium. I felt like I was going to pass out . . . soon after that, right after that, it just started shaking like crazy. The Bronco was off the ground completely, it was just rocking like this, and right soon after that the bank . . . dropped [the fault scarp formed]."

Near the Bronco, a Mrs. Knox was up a stream gulch facing a hillside looking for her husband (⊗ on map). She reported the formation of the fault scarp on the hillside directly as she watched, taking only a few seconds for the break to extend several miles along the range front. Mrs. Knox reported that the strong shaking from the earthquake snapped her head back and forth, so much so that she later had a neckache and headache—undoubtedly a severe case of whiplash.

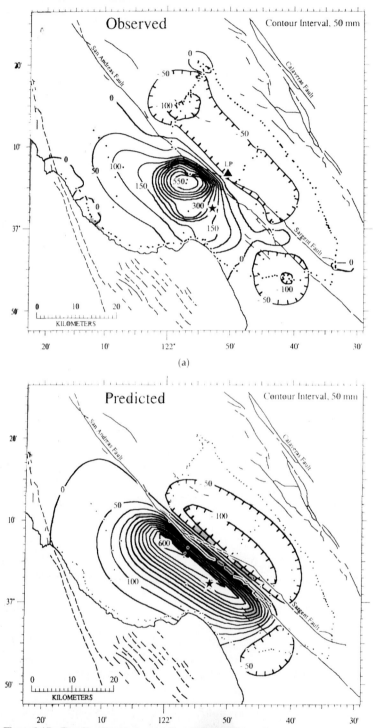

Figure 3.17   Results of geodetic modeling of the fault zone from observed changes in ground elevations (in millimeters) resulting from the 1989 Loma Prieta earthquake.

Figure 3.18   View of the fault scarp created in the 1887 Sonora, Mexico, earthquake.

fault scarps, whereas large shallow earthquakes with magnitude greater than 7 usually do. A controlling factor is depth of focus of the earthquake. Most earthquakes occur in the upper part of Earth's crust, at depths of 15 kilometers or less beneath the surface. At such depths earthquakes of about magnitude 6.5 or larger will usually have enough energy so that the fault motion reaches the surface, producing a scarp. There are a few exceptions. The 1925 Clarkston, Montana, earthquake produced no identifiable surface scarps. It had an estimated magnitude of 6.6 and a focal depth of 8 to 10 kilometers.

The fact that some larger earthquake faults do not always reach the surface and result in surface fault scarps is a particularly troubling characteristic as it is the larger damaging earthquakes that are of concern in earthquake hazards analysis. Faults that do not reach the surface are termed, appropriately enough, *blind* or *hidden faults*. Research on the subsurface geology and structure of the Los Angeles basin has revealed numerous potentially active hidden faults. Earthquakes produced by such faults could devastate the Los Angeles metropolitan area.

The 1994 Northridge earthquake in the Los Angeles urban area was on a blind thrust fault, the Pico fault. This fault was generally unknown to geologists and seismologists before the quake. The Northridge tremor was a magnitude-6.7 event, which originated at a depth of about 12 miles (19 kilometers). This earthquake was in many ways a duplicate of the 1971 San Fernando quake, as both were thrust-fault events, were the same size (Mw 6.7), and resulted in similar casualties. An important difference was that the San Fernando tremor ruptured the surface, while the Northridge event did not (Fig. 3.19).

The Loma Prieta earthquake of 1989, which struck central California, shaking much of the San Francisco Bay area, was magnitude 7.1, and was similar to the Northridge tremor in that it too produced no fault scarps. The rupture began at a depth of 19 kilometers and stopped at a depth of 3 kilometers. A study of data generated from this

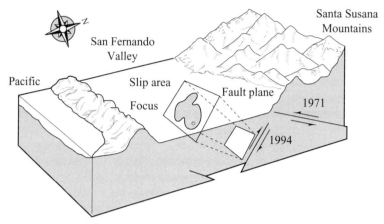

Figure 3.19   Thrust faults in the San Fernando Valley, which were active during the 1971 and 1994 earthquakes.

earthquake has allowed researchers to reconstruct the rupture process, including the time sequence development of the rupture front on the fault surface.

SUMMARY

Faults are fractures in which the blocks of rock move parallel to the fault surface. The study of faults involves understanding how and why movements occur. The rock adjacent to the San Andreas fault moves horizontally and parallel to the trend or strike of the fault. Thus, it is a strike-slip fault. This movement occurs because two blocks of Earth's crust slip past one another. Where Earth's crust is being compressed or shortened, movement may be directly up an inclined fault surface that is parallel to the fault dip. The movement is termed *dip slip* and the type of fault is a *thrust* or *reverse fault*. A second kind of dip-slip fault occurs where the crust is undergoing expansion. To make up for the extra volume due to the expansion of the crust, the hanging-wall block overlying the inclined fault surface will slip downwards. Such a fault is a *normal fault*.

Earthquakes occur along faults because of a buildup of stress in the rock. Stress builds as a result of friction on the fault surface, which resists any movement along it. Friction can commonly occur for two reasons: roughness of the fault surface and irregular geometry of the fault surface. Rough spots on fault surfaces are called *asperities*. They are barriers to movement and are an important element in the fault–earthquake process that is termed *stick-slip* or *unstable sliding*. Smooth and continuous movement may also occur with few or no earthquakes. This is called *stable sliding*, or *creep*. An example of creep occurs on the San Andreas fault, near Hollister, California. The San Andreas exhibits both stick-slip and creep types of behavior.

The study of fault scarps produced by large and damaging earthquakes has proven useful in understanding the earthquake process as well as in earthquake hazards studies. Fault scarps deform Earth's surface, and information about how much deformation has occurred is used to model the amount and direction of fault slip. Study of fault scarps that are produced in earthquakes and that are also recorded by seismographs has led to development of equations relating fault length to earthquake magnitude. These equations can be applied to calculate magnitudes for pre-instrumental earthquakes such as the 1887 Sonora, Mexico, event.

Our understanding of earthquake hazards has advanced much from study of the well-exposed San Andreas fault. However, other faults exist that are not exposed at the surface. Such blind or hidden faults exist beneath the surface of the Los Angeles basin and pose a great threat to the Los Angeles metropolitan area.

KEY WORDS

asperity	friction	steps
blind faults	Hollister	stick slip
Borah Peak	Lewis thrust	strike-slip
creep	normal fault	thrust fault
dip slip	Parkfield	Wasatch
Fort Tejon	San Andreas	

PART II:

EARTHQUAKE DATA ANALYSIS AND ITS CONTRIBUTIONS TO SCIENCE

CHAPTER 4

Earthquake Size and Location

INTRODUCTION

The development of successful seismographs toward the end of the nineteenth century was both an incredible achievement and an important advance in the study of earthquakes. For the first time it was possible to make a permanent record with time marks of ground movement in sequence that could be studied and analyzed. Such studies generated techniques that would allow estimation of the size and location of an earthquake.

EARTHQUAKE LOCATION

During the 1890s seismographs had been designed that could routinely detect and record waves from larger distant earthquakes (teleseisms). Owing to the efforts of John Milne, seismograph stations had been established by 1900 on all continents except Antarctica.

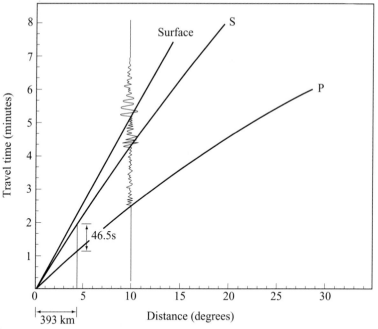

Figure 4.1   The travel-time chart with P-, S-, and surface-wave curves shown. This chart is useful in locating shallow earthquakes occurring in Earth's crust, those with depth of focus less than about 30 kilometers.

Sixteen stations were regularly sending records to England where the data from this worldwide network were being catalogued and evaluated. The collection of data allowed the construction of a basic earthquake location tool, namely the travel time chart. This chart is a plot of the travel time of seismic waves against distance from the earthquake (Fig. 4.1). Milne plotted the time of arrival of the phase of the maximum amplitude (usually the surface wave) at each station against the distance of the station from the known epicenter. The known epicenter location was usually determined from felt reports of observers. By 1899 Milne had plotted the P-wave arrivals as well, while in 1900 Richard Oldham was the first to plot and correctly identify the S-wave. Through use of the travel time chart, and records from at least three stations, it became possible to locate earthquakes by a method similar to triangulation in surveying.

The location technique is based on the fact that P- and S-waves travel at different velocities. Imagine the two waves as two cars in a highway race where the road is marked every mile. The cars pass the starting point (earthquake epicenter) at their full speed and at the same time. The P-wave car is traveling at 60 miles per hour and the S-car is traveling at 40 miles per hour. When the P-car passes the first mile marker after the start the S-car is one-third of a mile behind it. One hour after leaving the starting point the P-car is passing the 60-mile marker while the S-car is passing the 40 mile marker, 20 miles behind. Just as with the cars, there is a steadily widening gap between the P- and S-waves with time and distance from the epicenter. The S-car will reach the 60-mile mark 30 minutes after the P-car, whereas it was only 30 seconds behind the P-car at the

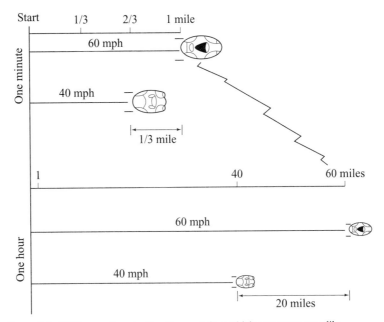

Figure 4.2   Difference in velocity of two moving vehicles creates a steadily widening arrival time gap at markers with distance from the starting line.

1-mile mark. Thus, knowing the arrival time difference, it is possible to determine the distance from the starting point (Fig. 4.2). In the automobile analogy the distance is known beforehand, but this is not the case with an earthquake.

The southern California earthquake of December 31, 1987, serves as an example of the determination of earthquake location by use of the travel time chart (Fig. 4.1). The raw data necessary to find the epicenter must come from the seismogram. Here the time marks are crucial. The recording drum for this seismogram was rotating at 60 millimeters per minute, so that 1 second equals 1 millimeter. Adjacent time marks are 10 millimeters or 10 seconds apart (Fig. 4.3). Note, reading with time increasing to the right, that the P-wave arrived at 21:35:06.0 Universal Time Coordinated (UTC). The S-wave arrived later at 21:35:52.5. This is a difference in S- and P-wave arrival time of

$$21:35:52.5 - 21:35:06.0 = 46.5 \text{ seconds}$$

The S-P time difference is sufficient to tell how far away the earthquake was from the station in Arizona. The difference can be obtained from the travel time curve (Fig. 4.1) where time is read from the vertical axis. There is only one distance from the earthquake epicenter where the S minus P time separation will be 46.5 seconds. It can be seen from Figure 4.1 that this distance is 393 kilometers. So the epicenter of the earthquake was nearly 400 kilometers from station WMZ. But in which direction? A circle with a radius of 393 kilometers can be drawn around station WMZ on a map. The epicenter is somewhere on this circle. The technique that can identify where on this circle the epicenter is located is sometimes called the *circle intersect method* and requires data from at least two other stations. Good, clear earthquake signatures were also obtained from:

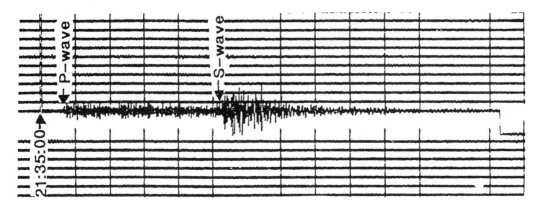

Figure 4.3   Seismogram of the December 31, 1987, earthquake in California as recorded at the Williams, Arizona, station.

	P-wave	S-wave	S-P time
PLM	21:34:17.9	21:34:31.0	13.1 seconds
BLP	21:35:05.0	21:35:48.6	43.6 seconds

Applying the S-P time difference to the travel time chart in Figure 4.1 it can be seen that PLM is 112.5 kilometers and BLP 371 kilometers from the epicenter. If circles are drawn with radii equal to these distances around the respective stations on a map (Fig. 4.4) all three circles intersect at only one point in southern California, the epicentral location. This is very close to a northerly trending group of faults along which the June 28, 1992, Landers, California, quake occurred. The Landers tremor was the largest earthquake in the United States in 40 years (magnitude 7.5). The December 31, 1987, earthquake represents stress release in the same area.

Obtaining a good location also depends on the number and distribution of stations with respect to the epicenter. Although it is possible to determine a location of an earthquake with only three stations, it is a case of more is better. Bad data from one station will have more of an impact where data from only three stations are available than where 10 are used. Station locations with respect to the epicenter are also critical. If stations lie all to one side of an epicenter, say to the west of it, a good location is not likely. The best situation would be to have stations arranged uniformly around the epicenter.

The most difficult unknown fact to determine in an earthquake location is the depth of the quake. The site of rupture initiation is located at depth and is termed the *focus* or *hypocenter*. If Earth were homogeneous, depth determination would be less of a problem. But Earth is not homogeneous with depth or laterally, i.e., parallel to Earth's surface. Earth consists of concentric shells or rock layers that vary in physical properties. Lateral rock property variations occur on all scales, from the change in rock type between ocean basins and continents, to that around salt domes (Fig. 4.5). The velocities, and therefore the transit times of the P- and S-waves, vary in the different rock types. The greater the variation in rock types, and the less well known the rock structure and layering, the more difficult it will be to use the travel time chart to locate an earthquake epicenter. This also makes it more difficult to estimate depth of focus. Note in Figure 4.5 that

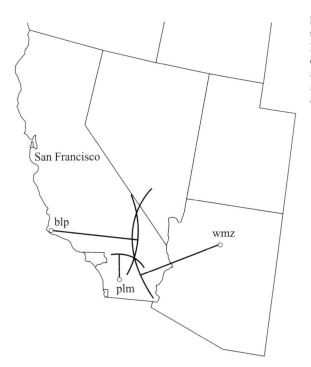

Figure 4.4 Location technique for the California earthquake of December 31, 1987. The three circles, represented in part by the arcs around the respective stations, all intersect at only one point, the estimated epicenter location.

to reach station S2, a wave traveling from the earthquake focus must travel through four rock types (R1–R4), whereas for a station at the epicenter the wave traverses only two. Rock changes that occur with depth are often better known than are lateral changes, so a station close to the epicenter will give a better depth estimate, because the ray path between focus and station will be steep.

The crustal model for eastern Turkey may be used as an example of its importance for location (Fig. 4.6). The crustal model contains three elements for each layer that directly affect the travel time and thus also location. These elements are layer thickness, layer velocities (P,S), and density of the rock (see Box 2.8). A change in any one or more of these elements will cause a change in travel time, which in turn will affect location.

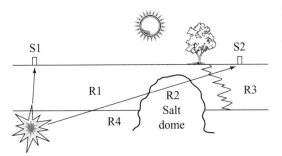

Figure 4.5 The effect of horizontal and vertical variations in rock structures on estimation of focal depth of earthquakes. The salt dome is an example of a naturally occurring body that can cause a change in lateral velocity of earthquake waves.

Depth (km)	Vp (km/s)	V_s (km/s)	Density
0-20	6.0 k/s	3.2 k/s	2.7 g/cm³
20-35	6.7 k/s	3.7 k/s	2.8 g/cm³
> 35	8.1 k/s	4.7 k/s	3.3 g/cm³

Figure 4.6 Crustal model for eastern Turkey. V_p and V_s are P-wave and S-wave velocities, respectively. This is a three-layer model with velocities and densities increasing downward, from layer to layer.

Finally, accurate estimate of origin time and travel time of the waves from the focus to station is especially important in determining depth of focus. A fairly large change in depth of focus causes a relatively small change in travel time to distant stations. Thus, depth is poorly constrained. Conversely, a station close to the epicenter will provide a better depth estimate because the travel time to it will show a larger change with a corresponding change in depth of focus. The rule of thumb often used is that an accurate estimate of the depth of focus may be obtained if the closest recording station to the epicenter is a distance equal to or less than the depth of focus.

Location of an earthquake focus is an inverse problem. Basically the solution to an inverse problem involves the interpretation of a result. This is in contrast to the direct problem where a result is the desired outcome. An example in everyday life might be a car engine. Let's suppose for the direct problem the goal is to find out what would happen if the clutch is let out too quickly before applying the gas. The result, of course, and the solution to the direct problem, is that the engine dies. The inverse problem would be to observe that the car engine dies and to find out why. The engine could have died because of the release of clutch, or it could be a lack of gas in the fuel tank, poor engine timing, vapor lock, or a number of other causes such as bad low-octane fuel. So in this case we hope to interpret a result correctly.

Inverse problems are common in earthquake studies. The location of a focus is an inverse problem because we are given the result, i.e., arrival times of P- and S-waves at stations, and the problem is the location of the source that will give those arrival times. The technique is to solve the inverse problem by using the direct or forward-modeling approach. This approach uses trial locations to see if a match can be obtained for arrival times. From felt reports or the circle-intersect technique, a trial location for the epicenter is usually available (Fig. 4.7). Assuming the arrival times from the station records are correct, and that the origin time of the earthquake derived from the travel-time curve is relatively accurate, then the travel times on the travel time chart are essentially constant for each station (stations A–D). For a fixed depth, the distance between epicenter and station is the only variable. If the distances are plotted on the travel time chart for the trial epicenter, data points for the stations do not plot on the P-wave curve. Thus, the trial location must not be correct. Also note that for station A, which lies south of the trial epicenter, the station is too far away from the epicenter for its data point to lie on the P-wave curve, whereas stations C and D to the north of the trial epicenter are too close to this trial location for their data points to lie on the curve. This indicates that the true position of the epicenter lies to the south of the trial location. Thus, by trial and error the actual epicenter location is estimated more accurately. The best epicenter location is the one that most nearly aligns all of the stations (A–D) with the curve. Usually a computer is used in such an evaluation as the computation can be quite time-consuming and tedious.

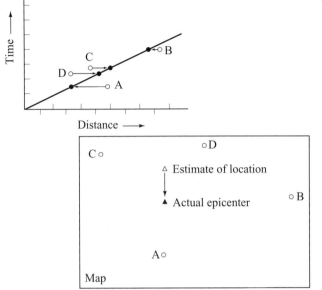

Figure 4.7 Use of the travel-time chart to estimate the location of an earthquake epicenter.

EARTHQUAKE DEPTH

Estimates of the depth of focus had been made as early as the Lisbon (1755) earthquake. John Michell estimated in a very general way that the depth of focus of the Lisbon earthquake "could not be much less than a mile or a mile and a half, and I think it is probable it did not exceed three miles (approximately 5 kilometers)." This is a rather remarkable estimate considering the information available to Michell. Most earthquakes occur at depths of less than 15 kilometers.

Robert Mallet, in his study of the 1857 Neapolitan earthquake, suggested a depth of focus of approximately 9 kilometers, again probably a rather good estimate. There are two factors that must be noted with respect to these early pre-instrumental estimates of Mallet and Michell. First, they are remarkably good estimates. However, this is not to be attributed to their methods of estimation or to a high quality of data available to them, but rather more to good intuition. Second, these early estimates more than likely had a strong influence on a view held widely for decades, namely that all earthquakes occur at rather shallow depth.

This was a notion that was challenged by H.H. Turner in 1922. Turner found from studies of compiled travel times of earthquakes that there were errors in arrival times that could be accounted for by deep-focus locations for some earthquakes. Final acceptance of the existence of deep earthquake foci came from an analysis published by K. Wadati (1928). Wadati examined earthquakes in Japan with the same epicenter but with different arrival times at the same stations (Fig. 4.8). Looking at curves drawn on a map for equal S-P times, a striking difference could be seen in spacing, which could only

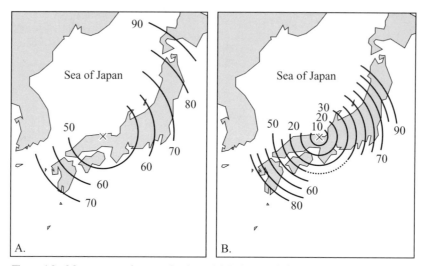

Figure 4.8 Map contours in seconds of equal S-P time for (A) deep and (B) shallow shocks in the Japanese Islands.

be explained by a difference in depth of focus. The broader-spaced contours represented a greater depth of focus.

Another tool can be used to estimate depth of focus. This is the arrival time of a P-wave that travels upward and reflects off Earth's surface in the area of the epicenter. This P-wave is known as pP (Fig. 4.9). The difference in arrival time between the direct P-wave and pP is a measure of the depth of focus.

Although most earthquakes occur in the mid to upper crust—that is, less than 15 kilometers in depth—in certain regions deeper earthquakes are more common. The deepest earthquakes, which have been widely recorded and thus have well-estimated depths, occur almost 700 kilometers beneath Earth's surface. Such earthquakes are usually not a threat because of their great distance from the surface, but they are an especially rich source of information about Earth's interior.

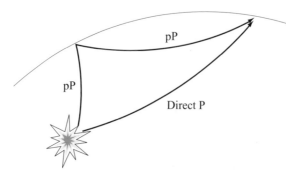

Figure 4.9 Depth estimation as based on the near epicentral reflection of P (pP). This path results in a travel-time difference between pP and direct P-waves.

An especially good example of a deep earthquake was the June 9, 1994, tremor that occurred beneath Bolivia at a depth of 600 kilometers. Because of its great depth and large size (magnitude 8.1) this was the first South American earthquake felt in North America, with reports of humanly perceptible ground movement as far north as Toronto, Canada. The timing was fortuitous as well, because when it occurred a large array of field seismographs was operating in Bolivia and Brazil, and so this was also the most completely recorded deep earthquake in history. The data collected allowed the discrimination in detail of deep-layer boundaries at depths of about 400 kilometers and 670 kilometers.

EARTHQUAKE SIZE

The earthquake magnitude is now a widely accepted concept in earthquake seismology, in structural engineering, and by the general public. It is, however, generally not well understood except by workers in a few scientific and engineering fields. The reason for the wide acceptance of the earthquake magnitude concept is its usefulness. The structural engineer uses magnitude and associated ground accelerations predicted from it as a way of estimating the level of ground shaking. This is useful in helping to determine the structural designs necessary to limit damage resulting from an earthquake. Magnitude is also helpful in earthquake hazards analysis where an important goal is to predict the potential damage of tremors in a geographic region. Earthquake magnitude is useful as well in generating data in long-term prediction of earthquakes and understanding strain in Earth's crust as a result of movement of the 100-kilometer-thick plates of crust and upper mantle (see Chap. 6).

The problem of determining the size of an earthquake has been a perplexing one, which was not altogether solved even with the advent of seismographs. In fact, the first widely accepted method of determining earthquake size from seismograms was not developed until the 1930s, some 50 years after the advent of successful seismograph systems. Like many advances in science, the groundwork is prepared first, and several researchers may be on the verge of discovery at the same time. This was the situation with the invention of calculus in the 1600s, when Leibniz in Germany and Newton in England were perfecting it simultaneously. This was also the case with earthquake magnitude with Wadati in Japan and Charles Richter in the United States, both developing techniques independently. Richter received the credit because his magnitude concept was published and widely distributed first. Richter himself gives credit to Wadati for having taken the first step (Richter, 1958, page 340, *Elementary Seismology*).

The approach used by Richter in developing a magnitude scale for earthquakes was to use amplitude of ground shaking in a new way. The principal problem in using the intensity of ground shaking as reported by observers is that it is often subjective data. However, the amplitude of ground motion as recorded on a seismogram is objective data. The only problem is how to use this as a measure of size, for several other factors are involved in determining the amplitude actually recorded on a seismogram at a seismic station. The most important factors affecting amplitude on a seismogram are amount of energy radiated, distance from the epicenter/focus, and magnification of the seismograph system. Richter was fortunate in that at the time of his development of a magnitude scale, southern California had a seismograph network operating with identical seismograph systems. This reduced the problem of development of a magnitude scale to one of accounting for the effects of distance on amplitude.

Richter had a great deal of data from southern California earthquakes recorded by the local network. His approach was to plot amplitude against distance for given earthquakes (Fig. 4.10). He knew that amplitude decreased with distance, but was surprised to find that for any two earthquakes of different size, this decrease was similar. That is, the curves for earthquakes that represent decrease of amplitude with distance were all roughly parallel to one another. Hence, the differences in amplitudes of any two earthquakes would be *independent* of distance, because the two amplitudes would always have a constant ratio. The only thing remaining then was to establish a baseline of measurement, or a zero-magnitude earthquake. This was done by accounting for the sensitivity of the instrument recording the data, the Wood-Anderson seismograph. The zero level was defined to be an amplitude of one-thousandth of a millimeter at a distance of 100 kilometers from the epicenter. The zero-level earthquake was chosen so that magnitudes of the smallest shocks that could be recorded would be a positive number. Over the years since then, however, instruments have improved in sensitivity so much that now earthquakes can be recorded that have negative magnitudes. A person with a sledge hammer near a sensor can generate a magnitude –4 event.

Richter selected the term *magnitude* to describe earthquake size, borrowing usage of the term from astronomy where it is used to describe the brightness of stars. Richter magnitude is based on the maximum amplitude of the largest waveform. For local earthquakes of moderate size recorded by the southern California network, this turned out to be the S-wave. Richter magnitude has also been called *local magnitude*, or M_L. This approach to estimate magnitude turned out to be unsatisfactory for large and distant earthquakes where the S-wave was not the most prominent phase. The best estimate of earthquake size must also be a good estimate of the energy released. Large and distant earthquakes usually had much greater amplitudes and thus much more energy associated with surface waves than with S-waves. So to estimate adequately the size of such an event an extension of the magnitude concept was necessary. This extension

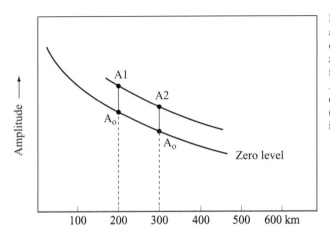

Figure 4.10   The distance-amplitude plot used for the determination of magnitude. The zero-level earthquake amplitude from a seismogram is A_0. Note that $A_1 - A_0 = A_2 - A_0$, so that the distance of the magnitude estimate (station) from the source is irrelevant.

of the concept to surface waves was carried out jointly by Richter and Beno Gutenberg at Caltech, and is termed *surface wave magnitude* (Ms).

Deep-focus distant earthquakes (teleseisms) posed yet another problem, for they do not develop good high-amplitude surface waves at the same period as do large shallow shocks. The solution was to base magnitude on the amplitude of the P-wave, which is not greatly affected by the focal depth of the source. Such a body-wave magnitude is termed Mb. These magnitude scales (M_L, Ms, Mb) gave fairly accurate estimates of size for many earthquakes, and for many years they were the standard in instrumental size estimates.

Magnitude as calculated by measuring the amplitude of different waves should be an average from a number of stations. As with location, the greater the number of stations used, the better the estimate. This is because the amplitude recorded on a seismogram is also a function of radiation direction, physical conditions along the wavepath, and ground conditions at the recording station.

The approach of measuring earthquake size by wave amplitude has its limiits for very large earthquakes (Ms≥7.3). The surface wave magnitude is inadequate to describe size for great earthquakes because magnitude is calculated from the amplitude of surface waves with periods near 20 seconds. The volume of rock required to produce a 20-second wave can be estimated as the cube of its wavelength (Box 2.6). The wavelength in this case would be about 60 kilometers, so the volume of rock would be about 216,000 cubic kilometers. But if the volume of rock generating seismic energy increases greatly, implying also a greater fault rupture length/area, even larger wavelength waves are generated that represent a significant amount of the energy released. This volume of rock and the energy released from it are not represented in the surface wave magnitude number. Thus for very large earthquakes the Ms scale reaches saturation.

The saturation problem of the Ms magnitude scale by very large earthquakes eventually led to the development of a new magnitude scale useful to describe the size of *all* earthquakes, the moment magnitude scale (Mw). An essential element in the calculation of the moment magnitude is the seismic moment (Mo). The seismic moment is an estimate of the size of the source and is defined as

$$Mo = \mu\,A\,u$$

where μ is the shearing strength of the rocks, A the area of the fault rupture, and u the average displacement on the fault. The seismic moment can be determined by observations in the field after the earthquake, or by the amplitude spectra (distribution/range) of seismic waves as determined from measurements made from earthquake records. The moment magnitude can be calculated from the seismic moment by

$$Mw = 2/3 \log 10\, Mo - 6.0$$

Very large earthquakes will saturate both the body wave and surface wave magnitude scales. These scales are not able to give an accurate estimate of size of very large earthquakes, especially events above Ms 8.0. This is analogous to velocity and an automobile speedometer. The speedometer may only record speeds up to 120 miles per hour, even though the vehicle may be able to go faster. A few examples of saturation of scales would be the very large 1960 Chile earthquake where Ms = 8.3, but Mw = 9.5. The 1964 Alaska earthquake, also unusually large, had an Ms = 8.4, and a better measure of size from moment magnitude of Mw = 9.2. In a practical sense, moment magnitudes cannot

BOX 4.1

Seismic Moment

The forces causing fault displacement and earthquakes can be described by using the concept of *moment* from physics. Recall the motion imparted to the atoms of a crystal lattice as an S-wave passes through rock (Fig. 2.18 and below). The force from the wave produces an up-and-down motion and results in a change in shape of the lattice termed *shear distortion*. The resistance to this distortion is known as the *shearing strength* (μ) of the material.

The change in shape of the lattice could be imagined as due to two equal but oppositely directed forces (F) some distance apart (d). The strength of the pair or couple of forces is the *moment* and is the product of the value of one of the forces times the distance between them: moment = force × distance = Fd. The greater F is, the greater the moment.

Seismic moment considers force in terms of work done along a fault surface. The measure is the area (A) of the fault surface, which ruptures in an earthquake, and the average amount of fault slip (μ). Force can also be measured by the amount required to overcome the strength of the rock (μ):

$$Mo = \mu Au$$

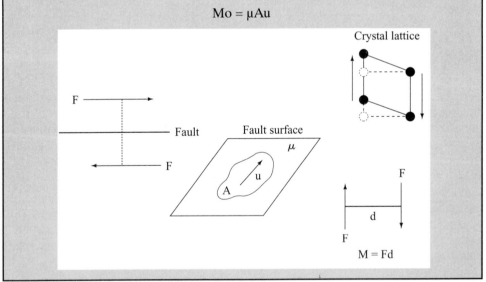

get much bigger than the 1960 Chile earthquake because there are very few active faults larger than the one that broke in 1960 (Box 4.1).

For smaller earthquakes, the difference in magnitude scales is not quite so large (Fig. 4.11). In any case, the moment magnitude is usually the better estimate of the energy released in an earthquake for earthquakes of all sizes (Fig. 4.11). Basically an increase of 1 magnitude unit is approximately equivalent to a 30-fold increase in released energy. Thus, the 1906 San Francisco earthquake estimated at Ms 8.25 was more than 30

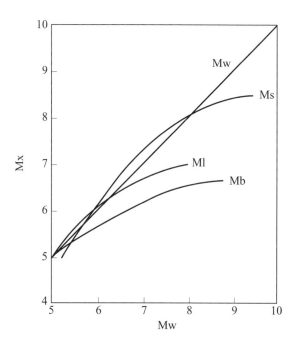

Figure 4.11 A comparison of Mw to the different estimates of earthquake magnitude presently in use. Mw = moment magnitude; Ms = surface-wave magnitude; Mb = body-wave (P-wave) magnitude; Ml = local or Richter (S-wave) magnitude. Mx = Mw, Ml, Ms or Mb.

times larger than the 1989 Loma Prieta earthquake of Ms 7.1, and roughly 900 times larger then the 1945 Hiroshima atomic bomb of about magnitude 6.0 (Fig. 4.12).

Although moment magnitude is coming into increasingly more common usage, for earthquakes of all sizes, the convenient and more familiar local magnitude (M_L) will continue to be widely used for smaller earthquakes by local/regional networks of seismographs for some time to come. The local magnitude is also a measure of frequencies generated by earthquakes, which are of interest to the structural engineer.

SUMMARY

The advent of successful seismograph systems made possible for the first time accurate estimates of location and size of earthquakes. Within the first 20 years of operation of pendulum seismograph stations, having principally John Milne design systems, sufficient earthquake records were available to begin to locate earthquakes.

The necessary first step in earthquake location was construction of the travel time chart. The chart is a plot of travel time of the P-, S-, and surface waves against distance. This had been accomplished by 1900. Because the waves travel at different velocities, the difference in arrival time for P- and S-waves could be used to estimate distance. Location could be estimated by drawing arcs, based on distance, around three stations. The arcs should intersect at the epicenter. This is the *arc* or *circle intersect* technique.

Most problems in seismology, such as earthquake location, are inverse problems. That is, such a problem is an interpretation of a result. The result could be earthquake arrival times. The interpretation would be the unique source location, which would give

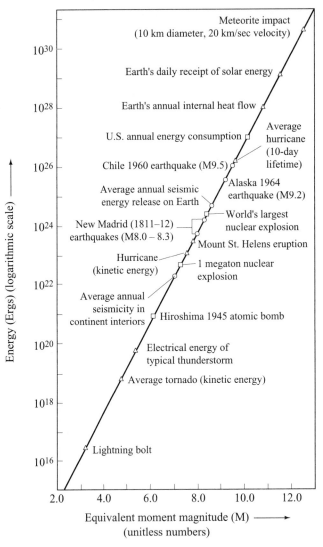

Figure 4.12 A plot of moment magnitude against energy (ergs) for comparative earthquakes and nonearthquake events.

the arrival times. The opposite problem is the direct or forward-modeling problem: Given a source location, arrival times can be calculated or predicted.

Of all source parameters, depth of focus is the most difficult to estimate. This is due to several factors. Variations in rock type and thus with seismic velocity affect location estimates, as do accurate estimates of travel time and origin time. The classical approach is to use P-waves from deep-focus earthquakes, which reflect from the underside of Earth's surface in the epicentral region (pP). Such phases are difficult to identify and measure for shallow events.

The earliest instrumental approach to determining earthquake size was developed by K. Wadati and Charles Richter. The Richter scale was based on the maximum S-wave amplitudes for southern California earthquakes. This concept was later extended to measurements of the amplitudes of surface waves (Ms) for large distant earthquakes, and P-waves (mb) for deep-focus earthquakes.

The best description of earthquake size, however, is based on the size of the source. The moment magnitude is an estimate of source size through seismic moment (Mo) and gives much better estimates of the size of the largest earthquakes. The Alaska earthquake (1964) had a much larger moment magnitude (Mw 9.2) than surface wave magnitude (Ms 8.4), although the difference in estimates for smaller earthquakes is generally much less.

KEY WORDS

body wave magnitude (mb)	inverse problem	seismic moment (Mo)
circle intersect	local Magnitude (M_L)	surface wave magnitude (Ms)
deep earthquakes	moment magnitude (Mw)	travel time chart
depth of focus	Oldham	Turner
direct problem	Richter	Wadati

The Earthquake Process

INTRODUCTION

The key breakthrough in understanding earthquakes had been the realization by the beginning of the twentieth century that a cause-and-effect relation existed between faults and earthquakes. This relationship was further documented and detailed by development of the elastic rebound theory as a result of the 1906 San Francisco earthquake. The Mino-Owari and San Francisco quakes had been natural laboratories for the study of the earthquake process, but, alas for science, all too infrequent an occurrence. Thousands of earthquakes occur throughout the world each year that could provide abundant information about the earthquake process, but they are usually too remote or too deep below Earth's surface for direct observation. A tool was needed that would allow such events to be analyzed. By the 1920s it was becoming clear that such a tool might have its basis in the earthquake seismogram itself.

AN IMPORTANT CLUE: FIRST MOTION OF THE GROUND

An important aspect of analysis of early records of earthquakes was the first arrival of the P-wave. The motion of the ground from the initial P-wave arrival, known as the *first motion*, can be either up or down as indicated by movement of a vertical component seismometer. Ideally, for an explosion, the ground motion would be up in response to the outward pulse of the compressive shock wave (Fig. 5.1). Thus a plot of first motion of P-wave arrivals on a map would ideally show this characteristic ground motion from an explosive source.

First-motion plots from earthquakes are not so simple, however, showing both up (compression) and down (expansion or dilatation) ground motions. In fact, these motions indicate a definite geographic pattern. Japanese and Italian workers recognized that lines could be drawn on a map of local and regional earthquakes separating regions of upward and downward first motion of the ground (Fig. 5.2). The first motions can be separated into quadrants or quarter spaces by two perpendicular lines on a map. The question naturally arose: Is this pattern a clue to the earthquake mechanism at the source? Or is it perhaps due to some effect of Earth along the wavepath as the wave travels through the rocks? This is a critical point, for our interest lies in the process at the source region. Early workers were taking a leap in faith in assuming that the first-motion arrival patterns represented motion in the source area. This is called the *assumption of conservation of phase signs* and implies that the transmission of the P-wave from the source to

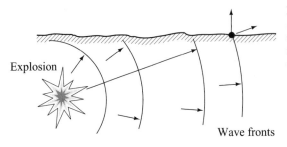

Figure 5.1 Shock-wave fronts advancing away from the explosive source. Earth's surface is forced to move up and away from the source explosion.

the station does not alter the first motion of the P-wave. The consistency of results using P-wave first arrivals suggests the conservation assumption is a reasonable one.

The next step was to compare the quadrantal first-motion pattern to the already existing concept that earthquakes and faults are related. Clearly a fault can be represented as a line on a map, so very likely one of the two perpendicular lines defining the quadrants represents the trace of the fault at the surface that is reponsible for the earthquake. But why two lines? And which line is the fault?

The theoretical basis for this phenomenon was a classical study relating earthquake mechanism (faulting) and first-motion patterns by H. Nakano (1923). Two models of the forces at the source were suggested: the single couple (Fig. 5.3A) and the double couple (Fig. 5.3B). Both models could produce the kind of ground motion pattern observed.

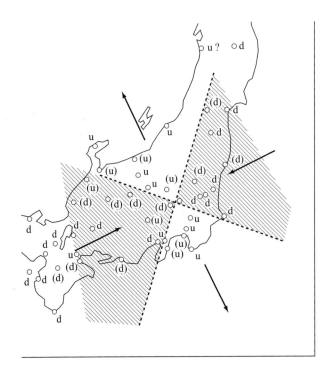

Figure 5.2 Sketch map of Japan showing compression (u) or upward ground movement from first motion of P-wave arrival at seismic stations. Downward (d) first motion is also shown. Note division of first-motion arrivals and ground movement into quadrants.

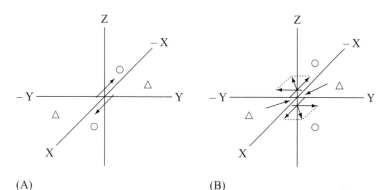

(A)   (B)

Figure 5.3　Models of forces at the earthquake focus: (A) single couple; (B) double couple. Arrows show direction of applied force. O = compression; Δ = dilatation. At Earth's surface, compression = up; dilatation = down first motion.

The single couple consists of two oppositely directed parallel forces separated by a perpendicular distance. Such a pair of forces is known as a *couple*. The separation of the forces produces a tendency for rotation, known as *moment*. The ground motion produced by the operation of such a force couple would be quadrantal, both in the source region and at Earth's surface (Fig. 5.4).

The double couple can be thought of as the addition of two perpendicular single couples. The result is a force system without rotation or moment. As with the single couple, operation of the double couple results in a quadrantal first-motion pattern at Earth's surface. A careful comparison of the elastic rebound theory (developed from an actual earthquake) to the two models favors the double-couple source mechanism, because of a lack of moment from the force couples.

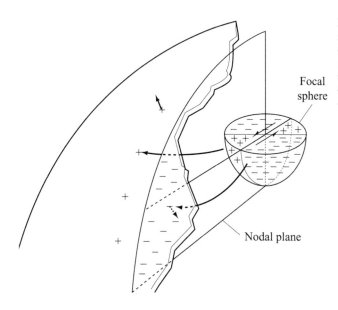

Figure 5.4　Distribution of first motion of the P-wave into quadrants, both in the source region (focal sphere) and at Earth's surface. Compression (+) and extension (−) or downward (focus directed) motion is indicated.

THE FAULT-PLANE SOLUTION: A MOST POWERFUL TOOL

The fact that the first-motion plot of P-wave arrivals on a map is a clue to the forces operating at an earthquake focus was an exciting and useful discovery. However, there are some problems with such map plots. First, which line represents the fault plane? This is a problem inherent in this type of analysis. Let's take another look at the Japanese earthquake in Figure 5.2. If the north-south line is chosen as the surface trace of the fault plane, then the ground motion pattern could be produced by a right lateral strike-slip fault. This is because rocks are compressed in the direction of motion on each side of such a fault (Fig. 5.5). However, if the east-west line is chosen as the fault-plane surface trace, the map pattern could result from motion on a left lateral strike-slip fault. The first-motion data alone cannot determine which line represents the fault plane. However, the *nature* or type of faulting is clear since the result in either case is compression released by strike-slip faulting. One interpretation often appealed to is to compare the trend of the two lines in the first-motion map to the trends of surface faulting in the area of the epicenter. Faults in the area that parallel a line on the map of first motions are likely candidates as earthquake sources.

A second problem with a map plot of first motions is that it is only valid for local/regional earthquakes where focus-to-station distances are relatively small. This is because seismic wave paths curve within Earth. This produces a distortion of the source first-motion pattern at Earth's surface (Fig. 5.6). If fault planes are chosen for distant earthquakes, based on the arrival pattern of first motions, the locations of these lines at the surface would differ from the correct projected traces.

The challenge then was to try to find a way to eliminate the problem of the curved seismic wavepath. Perry Byerly was the first to apply a projection technique to solve this problem. Although his approach, termed the *extended distance projection*, is no longer used today, it paved the way for a more easily visualized projection technique called the *stereographic projection*. Both of these techniques have the effect of elimi-

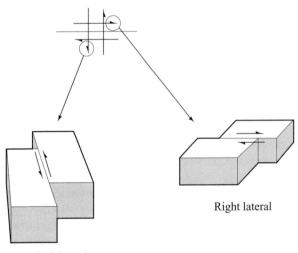

Figure 5.5   Quadrantal first-motion patterns yield two possible fault directions: east-west = left-lateral strike-slip fault; north-south = right-lateral strike-slip fault.

Right lateral

Left lateral

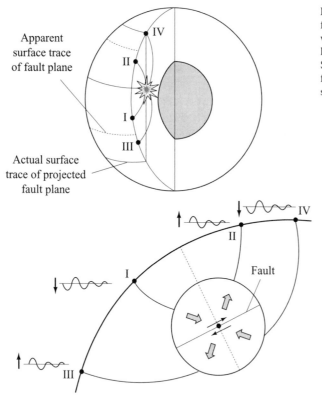

Apparent
surface trace
of fault plane

Actual surface
trace of projected
fault plane

Figure 5.6   Distortion of surface
first-motion pattern by curved
wavepaths. This can create false
location for nodal and fault planes.
Seismic stations (I–IV) shown with
first-motion directions at each
station as depicted by arrows.

nating curved wavepaths. The stereographic projection does this by considering only the region around the focus. This becomes possible by centering the plot in the source region rather than at Earth's surface. If the region around the focus included in the plot is sufficiently small then seismic wavepaths are short and straight. Instead of a map, the plot device is a sphere centered on the focus and is known as the *focal sphere*. The P-wave first ground motion at each station is plotted on the sphere's surface. This stereographic technique is a projection of the ground motion back along the wavepath to the focal sphere. Thus, the conservation of phase-signs assumption is critical, since it states that the direction of first motion does not change between the focus and Earth's surface.

P-wave arrivals at stations from distant earthquakes represent wavepaths that leave the focus traveling downward, so only the lower half or hemisphere of the focal sphere is necessary as a plot device (Fig. 5.7). This becomes transformed into a more manageable and useful two-dimensional representation of the hemisphere by the stereographic projection technique mentioned above. With this transformation, points and lines are projected onto a two-dimensional plot surface tangent to the bottom of the hemisphere (Fig. 5.7).

Because the focus of the earthquake lies at the center of the focal sphere, all wavepaths must also originate there, and the two perpendicular planes, known as *nodal planes*, one of which must be the fault plane, also center on the focus. All planes in the

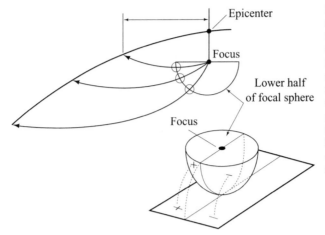

Figure 5.7   Wavepaths to stations reasonably distant from the focus (greater than 100 kilometers) leave the focus and pass through the lower half (hemisphere) of the focal sphere. Two-dimensional plots of first motions and nodal planes are traditionally made by projection to a plotting plane, which is tangent to the bottom of the lower hemisphere: (+) = compression; (–) = dilatation or focus-directed movement of rock.

corresponding stereographic projection are represented as great circle lines, or traces. This is similar to lines of geographic longitude on Earth's globe, which are also surface traces/lines that represent the intersection with the surface of Earth of planes that pass through the center of Earth.

The interpretation of the stereographic plot is straightforward. One needs only to remember that the view is a two-dimensional projection of a view downward into a bowl, or hemisphere. On the two-dimensional projection surface is seen the plot of points (station motions) and lines that represent nodal planes. The points are the intersection of P-wave paths with the surface of the hemisphere. Each point is represented by a symbol indicating either up (compression) or downward (dilatation) movement of the ground, which corresponds, respectively, to outward or inward movement of the focal region.

The lines in the projection of the surface of the focal sphere are the great-circle traces of the intersection of nodal planes with the sphere surface. One of these planes must be the fault plane, the other the so-called auxiliary plane, which is perpendicular to the fault plane. The stereographic projection of the focal region including the points of compression and dilatation and the nodal planes is known as the *fault-plane solution.*

An alternate representation of the fault-plane solution is the "beach ball" display. Instead of representing each station's motion at a point on the focal sphere, the quadrants created by the two perpendicular nodal planes are represented. By convention, the compressional quadrant is usually colored or shaded in, and the dilatational quadrant is left blank. This makes for a more visually effective diagram, especially when large numbers of stations are used in a solution, or when large numbers of fault-plane solutions must be displayed in a diagram.

Three kinds of beach balls will result if the fault motion is either pure dip slip, or pure strike slip. These three beach balls would then correspond to the three basic kinds of faulting: normal, reverse (or thrust), and strike slip (Fig. 5.8). Note that for normal faulting the central quadrant is unshaded, and P, the compressive axis, is vertical, as it represents the force of gravity. For the reverse or thrust-fault beach ball, the opposite is true: The central quadrant is shaded, and P is essentially horizontal. The strike-slip beach-

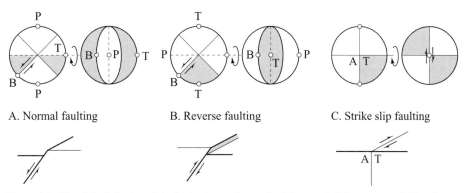

A. Normal faulting   B. Reverse faulting   C. Strike slip faulting

Figure 5.8   Beach-ball display of the fault-plane solution for (a) normal; (b) reverse; and (c) strike-slip faulting. Shaded quadrants represent compressional first motions. The right-hand ball of each pair is viewed with left ball rotated 90°; compressive quadrants are shaded. P = axis of compressive stress; T = axis of tensile stress. For A/T pair of right-hand diagram: A = motion away from viewer; T = motion toward viewer. The corresponding faults are show below each beach-ball pair.

ball pattern is seen as a distinctive checkerboard pattern. If slip is not exactly in the dip or strike directions, but at some angle in between, then a more complex-looking beach ball will result with both strike-slip and dip-slip components (Fig. 5.17).

The fault-plane solution technique has been a powerful tool ever since its inception in the 1920s in understanding earthquakes and faulting. The technique has been checked against earthquakes in which fault-slip directions and fault-plane orientations could be deduced by independent means, such as in cases where faults break the surface and displace the ground. Good solutions based on high-quality data are direct indicators of fault type (e.g., normal) and fault orientation.

The fault-plane solution allows analysis of faulting for earthquakes that are inaccessible: for example, for tremors occurring beneath ocean basins, in uninhabited areas, and for earthquakes that are deep enough or small enough to leave few or no surface indicators. The fault-plane solution was in fact a key element in gaining support for the plate tectonic hypothesis (see Chap. 6).

An example of the utility and power of the fault-plane solution method can be seen in its application to the study of reservoir water loads added to Earth's crust. The extra weight on the crust may increase earthquake hazard, as was suggested by the 1967 Ms 6.7 earthquake that occurred after filling of the Koyna reservoir in India. This was a nearly earthquake-free area before the reservoir was established.

Lake Mead on the Arizona-Nevada border behind Hoover Dam was likewise located in an area of few earthquakes. After reservoir filling, however, moderate and smaller-size earthquakes began to occur. Application of the fault-plane solution method to study the earthquake activity suggested a cause and allayed fears that a larger damaging event might be likely. Fault-plane solutions of the larger earthquakes indicated that stress was being released along north-south to northeast-southwest-oriented faults (Fig. 5.9). The results suggest release of tectonic stress stored in rock similar to the pattern in the area outside of the reservoir region and elsewhere in the western United States. Considering the lack of earthquake activity in pre-reservoir time it would appear that

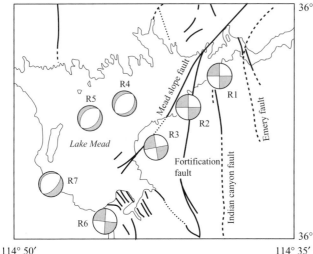

Figure 5.9   Fault-plane solutions for the Lake Mead area superimposed on a fault map. Note the agreement among the trend of the nodal planes of R1, R2, and the mapped faults.

reservoir loading of Earth's crust triggers the release of the stored tectonic stress. However, a decreasing rate of activity suggests a release of a significant portion of the stored stress and a likely decrease of the probability of a large event.

FOCUS VERSUS FAULT: EARTHQUAKE MODELING

The fault-plane solution is a powerful and still widely used tool for the analysis of earthquakes. But it does have its limitations. It has already been pointed out that this technique does not directly indicate which nodal plane is the fault plane. Nor does it reveal any of the other important details of the earthquake process such as rupture velocity and fault rupture length. This is because only a small portion of the seismic signature is used in the analysis, i.e., the first motion of the first P-wave arrival. These first arriving P-waves come from a very small region of the fault surface we call the focus, and represent only the moment of initiation of rupture. Details of the remainder of the rupture process remain hidden in the waveforms of the rest of the seismic signature.

The complex waveforms of the earthquake signature are generated by the earthquake process—that is, by the patterns of slip occurring on the fault surface. The problem then becomes, how do we analyze these waveforms to gain a better understanding of the earthquake process? Like many puzzles in earthquake seismology we are dealing with an inverse problem. That is, the earthquake signature on the record is a result, and the earthquake source process must be inferred by modeling the resulting signature.

Theoretical knowledge and data from earthquake sources have existed since the early days of earthquake recording: For example, two important concepts are the elastic rebound theory, and the Nakano force couple models. Also available since the late nineteenth century and early twentieth century has been the mathematical theory of

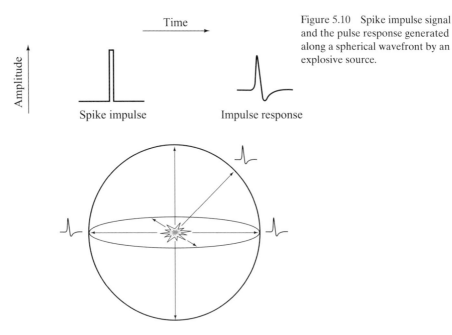

Figure 5.10   Spike impulse signal and the pulse response generated along a spherical wavefront by an explosive source.

waves in general and elastic or seismic waves in particular. This mathematical theory of waves has been applied to the development of additional models. For example, an underground explosion would ideally radiate a spherical pressure or shock wave outward from the focus or point of energy release. In the simplest case the pressure wave could be created almost instantaneously, with stress levels subsequently returning to normal. Such a source can be termed *impulsive* (Fig. 5.10). It follows then that it is possible to model the earthquake signature by using a combination of impulsive forces and to use this model to interpret the seismic source. There is yet another aspect of the seismic signature that must be considered if modeling is to succeed. This is the effect of Earth's structure and physical properties in the area of the earthquake source or focus, and the way in which it impacts the earthquake signature.

The ultimate goal of earthquake source analysis is to gain an understanding of the earthquake source process. To do this the area in the vicinity of the source cannot be ignored as it may modify the seismic waveforms and the resulting seismic records used in the source analysis. The kind of effect that the region around the source will have on the seismic waveform depends on the source structure. By source structure is meant the detailed layering in the focal region. Ideally, this would include the thickness and thus depth to rock layer boundaries, such as the depth to the crust–mantle boundary, as well as the velocity through the rocks contained in those layers (Fig. 5.11). This structure is important because it can change the resulting seismic signatures when the waves encounter the boundaries between these rock layers. This in turn will require a change in the impulse model to account for the changes at these boundaries. As an example, a one-layer case might be considered (Fig. 5.11). As the P-wave leaves the focus of an earthquake the expanding spherical wavefront includes an upward-traveling P-wave

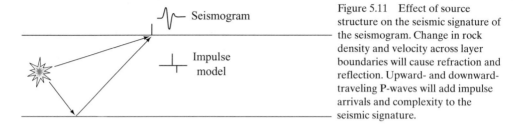

Figure 5.11 Effect of source structure on the seismic signature of the seismogram. Change in rock density and velocity across layer boundaries will cause refraction and reflection. Upward- and downward-traveling P-waves will add impulse arrivals and complexity to the seismic signature.

and a downward-traveling P-wave, both of which will encounter and bounce or reflect off boundaries. The way to include these effects in a model is to create model seismograms, known as *elementary seismograms*, which result from the individual reflections, and then to add their effects to obtain the resulting final model seismogram.

Although the structure of Earth in the area of the source will modify the waveforms and thus the signature on a record, what remains in the earthquake signature is the influence of the seismic source, which is ultimately what the research effort is focused on. Fortunately the seismic source has the strongest effect on the earthquake signature as it contributes to amplitude, separation of phases, and shape of waveforms.

The double-couple source mechanism has a radiation pattern for P-waves that varies with direction. The amplitude of a P-wave varies from zero, in the direction of the nodal planes, to a maximum at 45 degrees from the nodal plane direction. Thus, the focal mechanism and its orientation have a strong effect on the relative amplitudes of the P-waves recorded at seismic stations (Fig. 5.12). The focal mechanism also determines the polarity or first-motion direction of the P-wave.

Asymmetry in energy radiation patterns also has an effect on the amplitude radiation pattern of P-waves. An explosion (in the ideal case) would radiate equal amounts of energy in all directions. This is not the case with an earthquake. A greater amount of energy is radiated in the direction of propagation of the fault rupture. This results in an asymmetric radiation pattern with greater P-wave amplitudes in the rupture propagation direction (Fig. 5.13).

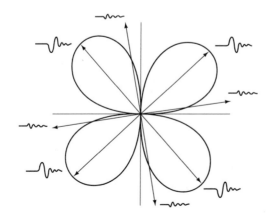

Figure 5.12 Effect of radiation direction with respect to fault-plane trend on amplitude of P-wave. At 45° from the fault direction, P-wave amplitude will be a maximum, decreasing to zero in the direction of the fault trend.

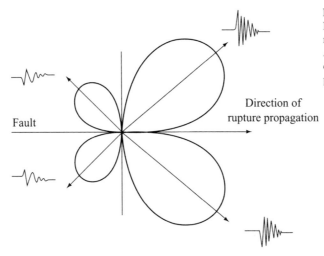

Figure 5.13   Asymmetric pattern of P-wave amplitudes seen in the radiation of earthquake waves. Amplitudes are higher in the direction of fault-rupture propagation.

Direction of rupture propagation

Fault

The depth of the earthquake source controls the separation of phases. This is because of different path lengths, e.g., P and pP (Fig. 5.14). Finally, the shape of the waveforms in an earthquake signature is controlled by the characteristic time history of release of energy. This is termed the *source-time function*. One example already given is the single instantaneous force impulse in an ideal explosive source (Fig. 5.11). The energy release is much more complex for earthquakes. Often energy release and fault slip will occur over a time period of more than 10 seconds in larger earthquakes. In fact, the release of energy in large earthquakes will frequently occur in distinct multiple events separated by several seconds. So the source-time function can be quite complicated (Fig. 5.15).

The modeling of so many different parameters as phase separation, waveform shape, and amplitude for a number of seismograms from different stations was nearly impossible before the advent of high-speed programmable computers. By the 1980s

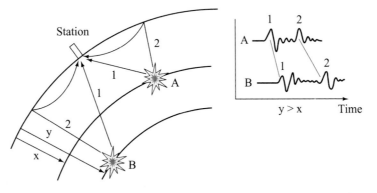

Figure 5.14   Effect of depth of focus on the separation of phases (waveforms) in the earthquake signature. Source *A*, which is shallower, has phases 1 and 2 closer together in arrival time than source *B*, which is deeper.

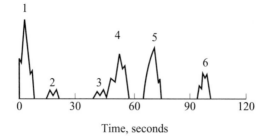

Figure 5.15   Source-time function for the 1967 Mudurnu Valley, Turkey, earthquake. The presence of six distinct rupture subevents are shown. The area under the curves provide the seismic moment (see Fig. 5.18).

both theory and technology had advanced sufficiently so that earthquake source modeling was relatively easy. Today, source modeling can be conducted on personal desktop and laptop computers.

The ability to interpret the results of earthquake source modeling is a function of the quality and type of data available. As always, the larger the data set, the better the chance of satisfactory analysis. Especially important is the period of the earthquake waves used. Resolution of finer details of the source process can be obtained by analysis of shorter period waves.

A magnitude-7.0 earthquake may rupture a surface of approximately 40 kilometers length extending to a depth of as much as 15 kilometers. Surface waves with a period of 200 seconds will have a wavelength of approximately 800 kilometers (Fig. 5.16). The source dimension is less than the wavelength and thus below the resolution possible with surface waves with a period of 200 seconds. The earthquake will appear to be

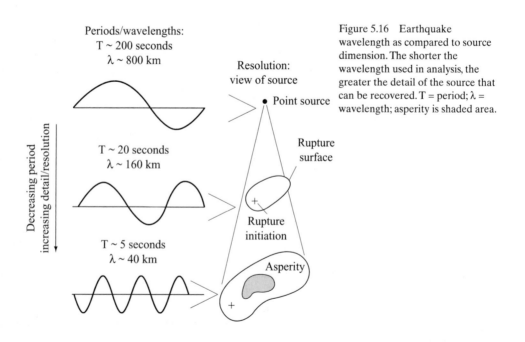

Figure 5.16   Earthquake wavelength as compared to source dimension. The shorter the wavelength used in analysis, the greater the detail of the source that can be recovered. T = period; λ = wavelength; asperity is shaded area.

a point source rather than a fault with respect to very long period waves. As a point source, surface wave analysis will yield seismic moment and fault orientation.

Perhaps most useful for analysis of larger earthquakes (Mw ≥ 7.0) are long period body waves, in the range of a few seconds to wave periods of dozens of seconds. This is because these waves have greater detailed resolving power than do surface waves and are still widely recorded at teleseismic distances. The shorter wavelengths (40 to 160 kilometers) are about the same scale as the rupture dimension on the fault surface, and analysis will reveal details such as duration of faulting, and the separation of individual events of slip (Fig. 5.16).

FAULT-PLANE SOLUTION VERSUS EARTHQUAKE MODELING

Earthquake modeling analysis is routinely applied to larger (Mw ≥ 6.0) earthquakes, as they are widely recorded. For smaller shocks, despite the limitations of the method, fault-plane solutions are still often the only tool available for source analysis due to the more restricted database. Because smaller earthquakes have a proportionately greater high-frequency content, and are well recorded closer to the focus, long-period wave analysis is not usually an option. Indeed, events Mw < 3.0 can only be well recorded by local networks. These tremors are termed *microearthquakes* and can reveal much about the tectonics of an area, especially if larger shocks are infrequent.

Northern Arizona is an area that has a low frequency of larger shocks. The largest earthquakes in the state occurred in 1906, 1910, and 1912, all about mb 6.0. These events occurred before the techniques of fault-plane solution had been developed and before sufficient seismic stations were operating to provide a database for source analysis. Nevertheless, smaller earthquakes occur more frequently and are well recorded by the local network. Arizona has a few M_L 3.0–4.0 earthquakes per year. These events are often well recorded out to about 1000 kilometers from the focus and are routinely analyzed using the fault-plane solution approach. Study of M_L 3.0–4.0 tremors between 1980 and 1989 in northern Arizona suggests extension of Earth's crust along generally northwest-southeast trending faults. This picture was confirmed by a rare larger event in 1993 of Mw 5.3, which generated enough data to allow both fault-plane solution and earthquake source modeling. The 1993 tremor occurred along a northwest-trending fault as well (Fig. 5.17).

THE HIDDEN IS REVEALED

The application of earthquake source modeling on high-speed computers has been a key development in increasing our understanding of earthquakes. This technique provides the capability of resolving such details as rupture velocity and direction, seismic moment, fault orientation, rupture dimensions, and rupture complexity, such as the presence of multiple events. Several case studies will illustrate the power and utility of earthquake source modeling.

The Anatolian region of Turkey is dominated by a large strike-slip fault, very similar in many ways to the San Andreas fault of California. Rocks on the north side of the

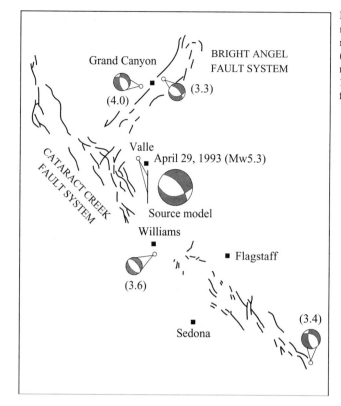

Figure 5.17 First-motion focal mechanism solutions of smaller northern Arizona earthquakes ($M_L \le 4.0$) compared to the source modeling of the moderate ($M_W 5.3$) 1993 Cataract Creek event and to fault traces in northern Arizona.

fault move east relative to those south of it (Fig. 5.18). This results in relatively frequent large earthquakes. The July 22, 1967, Mudurnu Valley earthquake was the same size as the San Andreas/Loma Prieta earthquake of 1989 (Ms 7.1). The Mudurnu Valley event has been a controversial earthquake because of different results obtained from several analyses, none of which completely explain the complexity of waveforms observed at seismograph stations.

Ali Pinar and colleagues applied earthquake modeling of P-waves to this earthquake with some interesting results. Their interpretation suggested that the complex waveforms were a reflection of complex source mechanics. In fact, waveform modeling suggested that the Mudurnu Valley source actually consisted of six events (Fig. 5.18). The process began with strike-directed slip near the center of an eventually 80-kilometer-long rupture zone. This event generated pressure at both ends of the rupture, resulting in two smaller events, one a normal fault, the other a reverse fault (Fig. 5.18, solutions 3 and 2). This was followed by three larger strike-slip events that migrated to the west, causing further slip along the larger rupture zone (Fig. 5.18, solutions 4, 5, and 6). Other details of source process, which may be deduced from earthquake source modeling, include the direction of rupture on the fault surface, the presence and distribution of asperities, and depth of focus.

Studies of the 1989 Loma Prieta earthquake provide interesting details about the source process. Waveform modeling from independent studies shows an uneven distri-

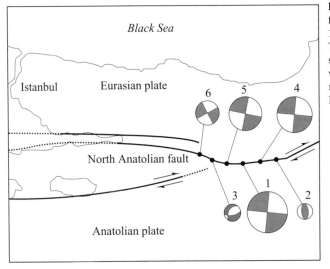

Figure 5.18   The North Anatolian fault of Turkey and the July 22, 1967, Mudurnu Valley earthquake. The numbers indicate six distinct subevents, in chronological order, which taken together represent the rupture process for the tremor (see Fig. 5.15).

bution of slip. The rupture direction was bilateral, i.e., the fault rupture proceeded from the focus in both directions along a fault strike away from the source, to the northwest and southeast. The rupture front propagated at about 80 percent of the S-wave velocity (3.0 km/sec). Most of the slip during the Loma Prieta earthquake occurred between a depth of 9 and 16 kilometers. The slip was concentrated in two main areas to the northwest and southeast of the focus (Fig. 5.19). These areas of high slip may represent asperities, but this is by no means certain.

Source modeling may also be applied to earthquake focal depth, which is at one and the same time a very important yet also a rather difficult parameter to estimate. This is because for larger shallow events there is often interference between phases arriving almost simultaneously, producing complex waveforms that are difficult to interpret. Body waveform source modeling is well suited to the interpretation of complex waveforms. A study of a large Indian Ocean earthquake by Seth Stein and colleagues serves as an example (Fig. 5.20). By matching the actual waveforms recorded with waveforms generated by modeling the source at several depths, it was found that a source located at a focal depth of 12 kilometers was the best match.

Understanding the earthquake process operating at the source is critical in areas where earthquake hazards are high. Source details such as accurate source depth, fault orientation, and knowledge of the rupture history—rupture direction, velocity, slip heterogeneity—are all useful results of modeling that can be applied to studying crustal deformation, as well as predicting ground shaking in urban areas and in improving building design to resist ground shaking. For example, buildings directly in the path of a propagating rupture experience more severe ground shaking and greater damage. This was particularly so in the case of the tremor that struck Kobe, Japan, in 1995. There the strike-slip fault and the rupture propagation along it were aimed like a dagger at the heart of Kobe. The results were disastrous: some five thousand people died and major damage was inflicted on the city.

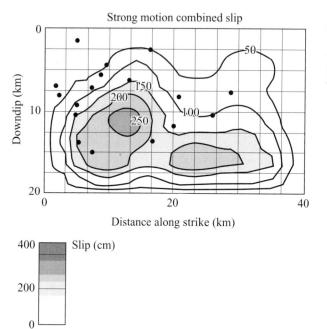

Strong motion combined slip

Downdip (km)

Distance along strike (km)

Slip (cm)

Figure 5.19 A contour diagram of slip amounts (centimeters) for the Loma Prieta earthquake as interpreted from seismic data. The fault plane is the plane of the diagram.

SUMMARY

Understanding of the earthquake source process has its roots in the discovery of the cause-and-effect link between faults and earthquakes. The elastic rebound theory provided the basis upon which focal mechanism theory could be built. The theoretical base was provided by Nakano's analysis of force couples at an earthquake focus, and it agreed well with first-motion ground patterns from earthquakes plotted on maps by early Japanese and Italian seismologists.

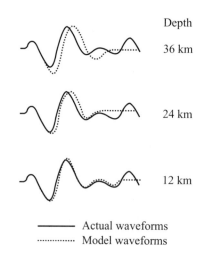

Depth

36 km

24 km

12 km

——— Actual waveforms
············· Model waveforms

Figure 5.20 An example of the use of waveform modeling to determine focal depth. A focal depth of 12 kilometers is the best match between synthetic seismograms and the actual earthquake signature.

First-motion plot of P-waves was the first tool to be applied to analysis of the earthquake source mechanism. Such plots, called fault-plane solutions, indicate the type of faulting (e.g., normal) at the focus and the orientation of two perpendicular planes, one of which is the fault plane. The fault-plane solution plot is a stereographic projection of a hemisphere centered at the focus upon which the potential fault planes appear as great-circle traces. A limitation of the fault-plane solution technique for P-waves is that it does not unambiguously allow one to choose which plane is the fault plane. Such choices are often inferred by comparing the results to regional fault trends and by looking at aftershock distributions, or by examining long-term geographic distribution of microearthquakes.

Analysis of the whole earthquake signature reveals much greater information about the earthquake process. Earthquake source modeling using computers has allowed detailed analysis of earthquake source mechanisms. Modeling of the complex waveforms with synthetic earthquake signatures has been successful in revealing depth of focus, rupture velocity and direction, area and distribution of slip on the fault surface, and resolution of multiple slip events (earthquakes) spaced closely together in space and time.

Thus, the earthquake source mechanism is no longer a complete mystery, and the results of source modeling efforts are being widely applied to studies of crustal deformation, earthquake hazards, and structural engineering.

KEY WORDS

bilateral slip	first motion	nodal plane
Byerly	focal sphere	single couple
compression	impulsive	source structure
dilatation	Loma Prieta	source-time function
double couple	moment	stereographic projection
elementary seismograms	Mudurnu Valley	
fault-plane solution	Nakano	

CHAPTER 6

Plate Tectonics

INTRODUCTION

Arguably the greatest contribution that the study of Earth has made to human thought has been the concept of plate tectonics. Like any important idea in science, plate tectonics is successful because it not only answers many existing questions about Earth, but also works well as a predictive tool. The clues leading to the plate tectonic hypothesis were in plain sight, but it took many years of research before the earth science community was ready to accept such a revolutionary concept. The study of earthquakes and earthquake source processes played a very important role in the acceptance of the idea of plate tectonics. But in the beginning of this story the groundwork had to be laid by the slow and laborious collection of data from many sources.

PUZZLES AND PIECES

The basic idea behind the plate tectonic hypothesis is that Earth's surface is mobile and in continuous slow motion. This concept is difficult for many to accept even today, but based on the evidence it seems an inevitable conclusion. For many who study Earth, a mobile surface seemed the only reasonable way to explain a number of observations, some of which were a paradox.

The earliest observation requiring explanation was the shape of coastlines. The Englishman Sir Francis Bacon upon seeing early maps noted the close fit of the coastlines of Africa and South America on opposite sides of the Atlantic (Fig. 6.1). By 1859 Antonio Snider-Pelligrini was suggesting, based on such observations, that the continents had moved. Snider-Pelligrini believed that the breakup of Africa from South America was due to instability produced when the continents were together on one side of Earth. This postulated instability was soon seen to be inadequate to move continents. Nevertheless, the fit of coastlines seemed too perfect to be a coincidence.

Events were coming to a head by the turn of the century. An American, F. B. Taylor, suggested a north to south movement of the continents, later facetiously termed by some as "continental drip." Subsequently in 1912, the banner of continental movement was raised by Alfred Wegener of Germany. Wegener's primary contribution was to assemble data from the geological sciences that would support continental movement, or continental drift as this concept became known.

Figure 6.1   The shape of the Atlantic coastlines of South America and Africa are identical to a high degree. The regions of similar rock types of the same age also line up well (stippled and lined patterns), as do the truncated mountain systems of the two continents (long, parallel lines).

THE APPEAL OF CONTINENTAL DRIFT: THE MOUNTAIN THAT CAME TO MOHAMMED

A number of observations have puzzled geologists and paleontologists for years, for which no reasonable explanation seemed available. This was because it was assumed that Earth had immobile continental landmasses.

The climates of ancient times made no sense in the context of fixed continents. It must be recognized that our present distribution of climates with respect to latitude is a consequence in large part of solar radiation. That is, warmer climates are found near the equator where larger amounts of solar radiation are received annually; colder climates are for the same reason near Earth's poles. Furthermore, such a general distribution has to have held throughout Earth's history. Any other distribution of climatic belts across Earth's surface would require a shifting of Earth's axis of rotation, which would be recorded as a cataclysmic event in Earth history. There is no compelling evidence of such an event.

Data on ancient climates produces a paradox when considered in the context of fixed climatic belts and immobile landmasses. Fossils from rocks recovered from Spitsbergen Island in the Arctic Ocean show that about 200 million years ago the island was covered with tropical plants. Yet oddly enough about 270 million years ago in central Africa much of the land surface was covered with glacial ice. Both locations had climates then that were much different from today. How to account for such an odd distribution of climatic belts? Some have suggested that instead of trying to shift Earth's axis of rotation that Earth's crust has shifted. Having the entire crust slip or move over Earth's interior would also be a cataclysmic event for which, once again, no overwhelming evidence exists, as well as no mechanism.

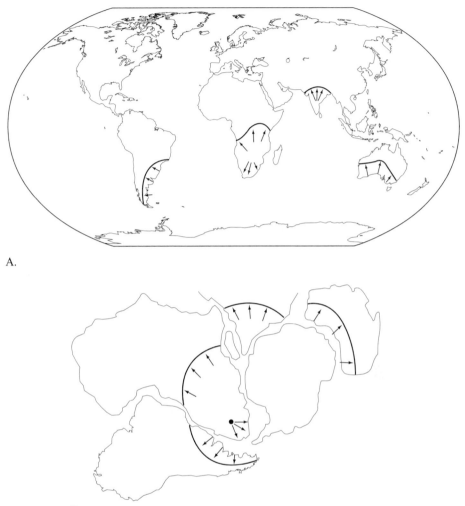

A.

B.

Figure 6.2   Locations and flow directions of ancient continental glaciers: (a) on present-day map of Earth, and (b) with reassembled continents.

The simple answer to the paradox of ancient climates embraced by Wegener is to move the continents and keep the climatic belts where they are. Such a scheme explains the climatic paradox as well as a number of other problems. For example, the central African glaciation was part of an ancient worldwide ice age that occured 270 million years ago (Fig. 6.2). As with our most recent ice age, glacial deposits were left on many continents. The last ice age clearly showed the control of presently existing climatic belts. That is, continental ice sheets were confined to more polar latitudes. There were no ice sheets at the equator, yet this is just where we find evidence for ice sheets of the ancient ice age.

Movement directions deduced from ancient glacial deposits suggest that some of these continental ice sheets had their source in the ocean basins. Such a situation is not possible. This would imply that the entire ocean basin would have to be filled with ice, which would require more water to be converted to ice by far than was the case in our most recent ice age. Wegener believed that the only reasonable answer to these problems of location and directions of ice flow was to reassemble the continents into one giant landmass, which he called Pangaea, centered around the South Pole. This arrangement eliminated the problem of oceanic glacial centers and equatorial continental ice sheets.

There was also other compelling evidence for continental drift. The southern continents all possessed fossil evidence of a widespread plant community termed the *Glossopteris* flora. This was a group of plants that was everywhere virtually identical and which existed at the same time in Australia, India, Africa, South America, and Antarctica. Continents separated by large ocean basins should develop, through evolution, their own distinctive plant and animal assemblage as, for example, Australia of today with its marsupial population.

Paleontologists, in answer to the Glossopteris problem, suggested the presence of ancient land bridges between the continents that would allow for the passage of plants and animals. This concept is not without some logic. During the most recent ice age, when sea level was 300 feet lower than today, a land bridge existed between Asia and North America across which animals and early humans migrated. When the glacial ice sheets melted, the bridge disappeared owing to rising sea level.

Land bridges between the Southern Hemisphere continents, however, would be impossible. Because they do not exist today they would somehow have had to disappear. A 300-foot change in sea level would not do the job, for the ocean basins are thousands of feet deep. The suggestion was that somehow these land bridges sank. This is physically impossible as land bridges, composed of lighter continental rocks, would be required to sink into heavier, denser oceanic rocks. This would be like expecting a cork to sink into water. Whatever happened to Atlantis, it did not sink beneath the waves.

A closer look at the fit of coastlines across the Atlantic between Africa and South America also lent support to mobile continents. If the fit of coastlines is to be more than coincidence then the rocks along this boundary on both sides will have to match. Detailed work by geologists over many years provided the evidence to support continental movement.

Broad areas of similar rock types and similar age known as geologic terranes exist in South America and Africa, which are separated by the Atlantic Ocean. These terranes fit together like pieces of a puzzle when the continents are juxtaposed (Fig. 6.1). Putting the two continents back together also seemed to solve the mystery of the truncated mountains. Mountain systems formed at the same time exist in South Africa north of Capetown and in South America south of Buenos Aires. Yet both mountain belts approach their coastlines at a high angle and end abruptly rather than gradually disappearing. When the continents are brought together the mountain belts line up like a sentence fitting back together on a torn page (Fig. 6.1).

The geologic evidence for continental drift seemed overwhelming to Wegener and many of the Southern Hemisphere geologists. Yet an important problem remained to be solved before the scientific community would accept this idea. What was the mechanism to move the enormous mass of the continents? What was the engine or power source?

THE ENGINE THAT COULDN'T

Alfred Wegener envisioned that as continents drifted they plowed through the rocks of the ocean floor, like giant ships at sea. The problem was why they should move at all. What force was responsible? The argument against continental drift revolved around this point. What forces were available to move continents, and how large must such forces be? Not much was known then about Earth's interior, and so the search had to be outwards.

Wegener suggested that two forces were responsible for continental drift. The first one was related to the tidal pull of the Sun and Moon on the rocks of Earth's crust. Here the idea was that since tidal pull may be gradually slowing the speed of Earth's rotation, it was also possible that the tidal pull might drag the crust westward.

The second force proposed as responsible for continental drift was Earth's rotation. This was based on the fact that as the planet rotates, there is an outward force generated that opposes gravity. This is the same force experienced by a stone tied to the end of a string and rotated by hand above a person's head. Without this centrifugal force the string will not remain taut, and the stone will fall. Centrifugal force increases as you approach Earth's equator, with the increase in force analogous to a longer string. The net effect of the centrifugal force and the force of gravity is to produce an overall inward attraction on Earth, which is directed to near Earth's center (Fig. 6.3). The contintental rocks affected by this force also oppose it, because as lighter crustal rocks they are more buoyant than the material beneath them. When a buoyant force is added to the inward force, it produces an overall force directed toward the equator, the pole-fleeing force or *polfluchkraft* of the Hungarian scientist Roland Eotvos.

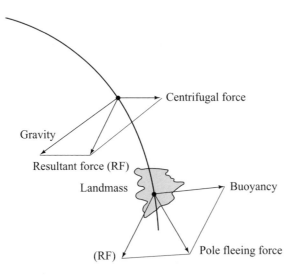

Figure 6.3   Eotvos pole-fleeing force. The forces on a continental landmass result in an equatorial directed force (pole-fleeing force).

Were these forces then adequate to move large continental landmasses through the ocean floor? Such a question can be answered by computation. Several scientists showed that neither force was capable of moving continents, and were in fact much too feeble. Thus, the continental drift adherents were left without an engine. They were also left without a leader: In 1930, Alfred Wegener was lost on an expedition in Greenland.

CONVECTION AND A MOBILE SEAFLOOR

The engine powerful enough to move continents is rooted in the concept of convection. Thermal convection occurs whenever water is heated to boiling in a pot. The water at the bottom of the pot adjacent to the heat source warms and expands, becoming lighter. As it does so it rises to the surface, carrying the heat energy with it. Near the surface adjacent to the cool air, the water cools and becomes denser, and then sinks to the bottom to repeat an endless cycle. In this fashion the movement of heated material transfers the heat energy, the basis of thermal convection.

Such a process was visualized for Earth's interior when the phenomenon of radioactivity was discovered. Because radioactivity generates heat, and Earth's interior contains rocks with radioactive minerals, the possibility was raised that Earth's interior was actually a heat source. Combined with heat from Earth's core, enough heat is available to make thermal convection a possibility.

Arthur Holmes was the first to suggest that thermal convection might provide the mechanism for continental drift. Holmes also put forward the idea that, instead of sailing through the ocean floor like ships through a sea, the continents are carried along like rafts in a passive fashion on thicker plates of rock (Fig. 6.4). This was one of the intuitive leaps of science, for there was little data and no proof to support his idea. Like Wegener's concept of continental drift, however, it did solve existing problems and explained some geologic enigmas.

Holmes's concept suggested that at the point where convection currents approach the surface they are forced to move sideways. As they do, they exert traction, or pull, eventually moving the continental landmasses sideways (Fig. 6.4). At the point where the convection current turns down, at an oceanic trench, the continent is dragged downward and folded. When convection stops, the buoyant continental rock rises to form mountains.

This concept by Holmes, although based on little data, was a remarkable insight. However, it raised some questions. Although convection was well accepted in liquids, what about within Earth? The passage of seismic waves through this region was proof that this material was solid. Could thermal convection occur in otherwise solid materials?

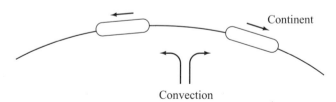

Figure 6.4   Arthur Holmes's concept on the relationship between convection and continental drift. The raftlike continents are pulled along by convection currents.

Although the idea of thermal convection in solids seemed a promising source of energy needed in the continental-drift hypothesis, it was not well received. The slow movement of solid mantle material, termed *creep*, was not thought possible by geophysicists of the time. The concept of thermal convection would require new supporting data, which strangely enough would come from technical advances resulting from World War II.

Acceptance of Holmes's concept of conveyor-belt tectonics and solid-mantle convection required new independent data. This came from the seafloor about which little was known before 1945. Two technical advances from submarine warfare in World War II were crucial for obtaining data on the seafloor. To detect submarines, depth-sounding equipment known as *sonar* had been developed. Sonar sends out a wave that will bounce off any object and return to the source. The travel time of the wave can be converted to water depth. The second technological advance was the development of magnetic sensors called *magnetometers*. Submarine hulls are made of iron alloys that have magnetic properties.

After World War II both depth-sounding and magnetic exploration were applied to the seafloor of the world's oceans, and the results were startling. Oceanographic exploration of the seafloor with depth-sounding equipment resulted in the first detailed topographic maps of the ocean bottom. The maps showed that the ocean floor was far from smooth, having a topography as varied as that of the continents. A 40,000-kilometer-long volcanic mountain chain, thousands of feet high, rose from the oceanic floor to girdle the globe. These mountains were designated the *oceanic ridge system*. Equally stunning were deep chasms or canyons in the ocean floor that became known as *oceanic trenches*. The bottoms of the deepest trenches are farther below sea level than the top of Mount Everest is above it. The magnitude and sheer size of the ridges and trenches demanded an explanation. Why did they exist? How did they form? What function did they serve? The critical clue in answering these questions would come from magnetic mapping of the seafloor.

MAGNETS, POLES, AND SUBMARINES: THE GREAT DISCOVERY

The breakthrough to support Arthur Holmes's concept of a moving seafloor would come from the mariner's best friend, Earth's magnetic field. The equipment developed to detect submarines would now be applied to map the seafloor of the world's oceans.

The seafloor is composed of a volcanic rock called *basalt*. Basalt contains magnetized iron compounds that acquire their magnetic properties as they form during cooling. The iron compounds will be magnetized in the same direction as any existing magnetic field as the basalt lava cools through a temperature known as the *Curie* temperature, at which point compounds are cool enough to acquire magnetic properties. The magnetism in basalts is one produced by the presence of Earth's coexisting magnetic field. Thus, if a basalt forms today from cooling of a lava flow, we would expect that if we place a compass next to it that it would continue to point to Earth's magnetic north.

Earth's magnetic field is very similar to that of a dipole or bar magnet. Any bar magnet will have two poles (dipolar), north and south magnetic poles. A compass needle is also a small dipole magnet, and so it lines up in Earth's larger dipole field (Fig. 6.5). The south pole of the needle will point to Earth's north magnetic pole, and vice versa.

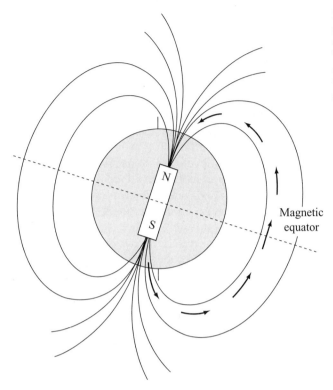

Figure 6.5 Earth's dipole magnetic field. The lines show the direction of the magnetic field. Thus, any magnetic object such as a compass would have to line up within this field, pointing toward a magnetic pole.

The south pole of the needle is called its north-seeking pole. The pole lines up along a direction parallel to a magnetic line of force. These lines of force either converge toward or diverge from the magnetic poles and define the direction of a magnetic field. For example, wherever the compass needle is in Earth's magnetic field, it is oriented parallel to one of these invisible lines of force. At Earth's magnetic north pole, the needle would point straight down into the ground. So to describe a magnetic orientation we must talk about the inclination as well as the horizontal direction. This is not obvious with a compass needle. Because only a horizontal direction is desired, the needle is balanced on its shaft so that it won't rotate about a vertical axis and become inclined.

Because basalts lock in the direction of Earth's magnetic field at the time they form, they preserve information about the past direction of Earth's magnetic field. Basalts can be found today that we know were formed millions and even billions of years ago. Interestingly enough, some of these basalts are magnetized in a direction exactly opposite to that of Earth's magnetic field today. It appears that at the time these basalts formed, Earth's poles had switched places or swapped, a phenomena now recognized as *polar reversal*. This has occurred many times in Earth's history, and it has an interesting effect on Earth's magnetic field strength.

The primary goal of the mapping of the seafloor's magnetism was to determine variations in magnetic field strength. The magnetometers to be used for this project measured the total magnetic field, including Earth's field and the field from the iron compounds in the ocean-floor rocks. The total magnetic field from these two sources varied in some unforeseen ways.

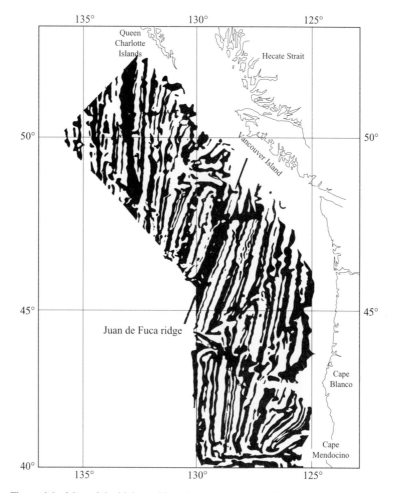

Figure 6.6   Map of the highs and lows in the magnetic field of the ocean floor.
Note that the pattern is parallel to the axis of the Juan de Fuca oceanic ridge (line).
Black stripes are highs in the magnetic field, alternating with white lows (Fig. 6.7).

Early airborne surveys across the oceanic ridges revealed a strange and symmet-
ric pattern (Fig. 6.6). This was a series of alternating highs and lows in the total field that
were symmetrically arranged about the ridge axis. The same type of pattern was seen no
matter where along the ridge the field was examined, or which ridge was surveyed. These
patterns were examined by F. J. Vine and D. H. Matthews at Cambridge University. They
recalled Holmes's theory of convection and made a link. Realizing that basalts were
forming at the oceanic ridges Vine and Matthews suggested that the seafloor was like a
gigantic magnetic tape playing out from the ridges and recording the direction of Earth's
magnetic field. They adopted the idea of H. H. Hess at Princeton, that the seafloor originated at the oceanic ridges and then spread away from it, a mobile seafloor. At the
ridge, the total magnetic field is a large value because as measured it is the sum of both
the present-day field and the magnetism of the seafloor rocks, which is added to Earth's

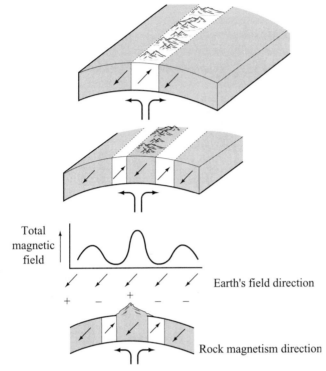

Figure 6.7   The relationship of highs and lows in the magnetic field across oceanic ridges to the magnetization of seafloor basalts. Dark areas are highs in the magnetic field. Highs occur where Earth's field and rock magnetism directions are the same.

field since they are both in the same direction (Fig. 6.7). Rocks on both sides of a ridge, however, which had been at the ridge in the past, had formed when Earth's magnetic field and its poles were reversed. Thus, today the rock magnetism of the seafloor rocks opposes Earth's present field direction, a situation analogous to the opposite poles of two-bar magnets, which repel. The magnetic field of the oppositely magnetized rock would actually subtract from the total field value detected by a magnetometer, producing lower values on either side of the ridge. This hypothesis was a major breakthrough, as it not only explained a puzzling pattern in the magnetic field but also provided a link to the Holmes and Hess convection-based motion, an engine to move the continents on plates of oceanic rock. Continental drift was back in a new guise, and the key support came from Earth's magnetic field that led in turn to the seafloor-spreading hypothesis. By the early to mid-1960s this had thrown earth scientists into open debate at all levels. The search now began for independent corroborating evidence to support seafloor spreading.

EARTHQUAKES AND PLATE TECTONICS

The idea of a mobile Earth's surface consisting of rock plates driven by convection would come to be called *plate tectonics*. Tectonics refers to the study of the deformation of Earth's crust and to the forces responsible for that deformation (Box 6.1). According to the plate tectonic theory, those forces are transmitted to the plate boundaries by the motion of the surface plates. For example, most of the United States lies on the North American plate, which is moving relatively westward, away from the spreading

BOX 6.1

Plate Tectonics and Earthquake Distribution

The majority of earthquakes occurring each year are located in relatively narrow bands that coincide with plate tectonic boundaries (stippled pattern). In fact, most plate tectonic boundaries could be located by these epicentral bands. The movement of plates against or away from one another generates forces that result in earthquakes.

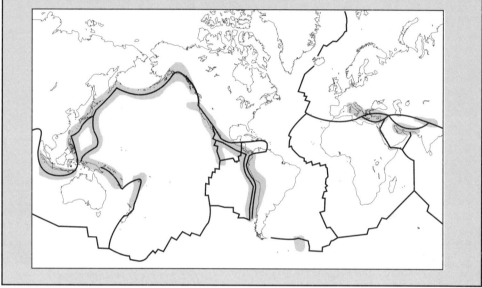

seafloor at the Atlantic oceanic ridge. This forces the North American plate against its western boundary, which in California is formed by the San Andreas fault. The adjacent plate to the west, the Pacific plate, should be moving to the northwest away from an oceanic ridge in the eastern Pacific. The motions of these two adjacent plates are transmitted to their common boundary, the San Andreas fault. If plate tectonic theory is valid, then the San Andreas fault should have right lateral strike-slip motion, which it does. This was clearly demonstrated in slip resulting from the great 1906 San Francisco earthquake. Supporting evidence comes from the application of the fault-plane solution technique. Most earthquake fault-plane solutions along the San Andreas fault and its branches show strong right lateral slip (Fig. 6.8).

So an important test of the plate tectonic hypothesis would be to determine the motion of the plates at their boundaries. Independent evidence of this motion would come from the study of earthquakes at those boundaries through application of the fault-plane solution technique. One of the first breakthroughs was the determination of plate motion at the relatively inaccessible oceanic ridges. Such a test would either confirm or eliminate the seafloor-spreading concept, the engine of plate tectonics.

The oceanic ridges had been a puzzle ever since the ocean mapping program had revealed their nature. Not only was there a 40,000-kilometer-long system of volcanic

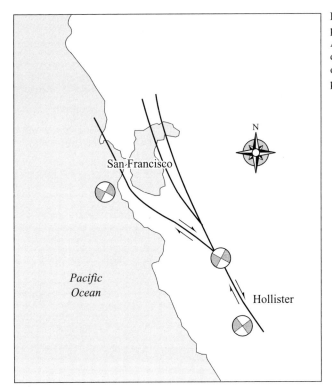

Figure 6.8 Representative fault-plane solutions along the San Andreas fault system. Beach-ball displays show right-lateral motion of the Pacific and North American plates.

mountains on the seafloor, but the ridge was divided into segments by large fracture systems perpendicular to it (Fig. 6.9). Initially it was believed that the fractures were strike-slip faults showing the relative directions of ridge offset by fault slip. However, by 1965 an interesting hypothesis consistent with seafloor spreading had been proposed by J. Tuzo Wilson of Toronto University. Wilson suggested that the fracture systems and the positions of the ridge segments had been inherited from the time of their formation and that the fracture zones would show slip related to spreading of the seafloor on either side from adjacent ridge segments. This motion would be the opposite of that expected if these fractures were merely strike-slip faults, which moved ridge segments away from one another (Fig. 6.9).

The application of the fault-plane solution technique to this hypothesis was not long in coming. Lynn Sykes of Columbia University assembled data from a number of earthquakes occurring along oceanic ridge boundaries throughout the world and found the results to be consistent with seafloor spreading.

The oceanic ridges had some earthquakes located along the ridge axis. These earthquakes invariably showed normal fault solutions, which is a confirmation of seafloor spreading and plate tectonics. Normal faulting accommodates the stretching or lengthening of Earth's crust. This suggested that the motion of the seafloor at the ridges was away from the ridge axis on both sides (A in Fig. 6.10). The foci of these earthquakes were also relatively shallow, indicating that the brittle rock of Earth's crust does not extend

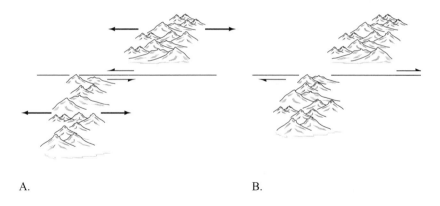

A.   B.

Figure 6.9   Oceanic ridge fracture zones had two intrepretations: (a) seafloor motion on both sides of the fracture; (b) simple strike-slip offset of the ridge.

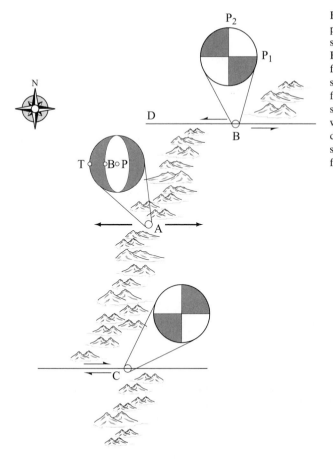

Figure 6.10   First-motion fault-plane solutions as evidence for seafloor spreading at oceanic ridges. Beach-ball displays show normal faulting along ridge axis (A), and strike-slip motion along ridge fractures (B,C). Arrows show seafloor spreading from ridges, while half-arrows show the direction of shear movement from spreading on both sides of the fracture zone.

very far below the surface. This would make sense if Earth's crust were warmer at the ridges, as could be attested to by volcanic activity and the presence of rising lava.

Epicenter locations along the fracture zones between ridge segments provided further support for seafloor spreading. The epicenters lie almost without exception between ridge segments (*B* and *C* in Fig. 6.10). This cannot be expected if the fracture zones are strike-slip faults responsible for offset of the ridge segments, but should be the case if slip along the fractures is the result of seafloor motion. Beyond the ridge segments (*D* in Fig. 6.10), the seafloor on both sides of a fracture is moving in the same direction, while between ridge segments the seafloor is moving in opposite directions, generating fault slip and earthquakes (*B* and *C* in Fig. 6.10).

The motion along the oceanic ridge fracture zones was confirmed by fault-plane solutions. The solutions showed the motion that is to be expected from seafloor spreading, the opposite of that which would occur from strike-slip separation of ridge segments. These new types of plate boundary faults are a special case of strike-slip faults and are termed *transform faults*.

The final piece of the puzzle would come from the oceanic trenches. According to plate tectonic theory this is the place where the plates converge. One of the plates is forced into another as it is *subducted* into Earth (Box 6.2). The geographic expression of this is a linear depression or trench where the plate bends as it subducts. All trenches worldwide were studied using the fault-plane solution technique. The Tonga trench of the southern Pacific serves as an example.

The Indian-Australian and Pacific plates converge at the Tonga trench. The rate of convergence is about 7 centimeters annually, which is among the higher rates. The result is a very seismically active boundary. The plot of earthquake foci indicates that a brittle Pacific plate is being forced west and underneath the Indian-Australian plate (Fig. 6.11). To the east of the plate boundary, earthquakes on the Pacific plate show extension-related normal faulting (*B* in Fig. 6.11). This may seem confusing at first until it is realized that at this location the Pacific plate is bending as it is forced downward at the trench site. This bending action stretches the upper surface of the plate, resulting in normal fault slip and earthquakes. Thrust faulting is seen beneath the trench as the two plates are forced together at the subduction site and move past one another (*A* in Fig. 6.11). The downward bending of the Pacific plate causes a more complex response along the transform fault, which terminates the trench to the north (*C* and *D* in Fig. 6.11).

The extensive study of plate motions at plate boundaries by the application of fault-plane solution techniques has confirmed in great detail the postulated motions of plate tectonics. This result provides very convincing independent evidence in support of the plate tectonic hypothesis.

SUMMARY

Analysis of data from the study of earthquakes played a key role in the acceptance of the concept of a mobile Earth crust by the scientific community. Data in support of crustal motion came first from the study of the geology of the continents.

BOX 6.2

The Shadow of Subduction

Subduction zones had been a puzzle years before plate tectonics was recognized. They had been termed Benioff zones after Hugo Benioff at Caltech who was the first to call attention to the fact that they were the site of a dipping band of earthquake foci, extending into Earth. The cold plate interior, in contrast to the warmer mantle around it, supports brittle fracture and faulting, thus generating earthquakes at greater depths than otherwise possible. The earthquake foci appear to produce a shadow of the subducted plate.

Temperature differences do not explain the deeper earthquakes, however, as pressures are too high for brittle fracture. High-pressure fault slip appears to be a result of the change of mineral structure under pressure at great depth.

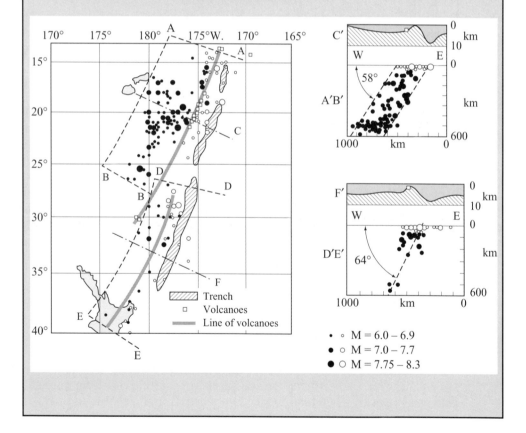

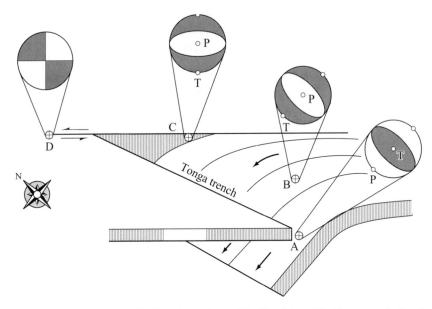

Figure 6.11   Convergence of Indian–Australian and Pacific plates at the Tonga trench. Beach-
ball displays of fault-plane solutions lend support to underthrusting by Pacific plate.

A number of different workers had recognized that there were problems that
could not be explained if the continents were fixed in position. These included problems related to distribution of ancient fauna, matching continental coastlines, ancient
glacial patterns, and truncated mountain systems. German scientist Alfred Wegener assembled information about these geologic puzzles. He found that the puzzles could be
explained if continents were able to move, a concept known as continental drift.

Although continental drift was successful in explaining a number of puzzling geological problems, its primary drawback was a source of energy sufficient to move large
continental landmasses. As little was known about Earth's interior at the time, external
forces were the bases of proposed energy sources. Both tidal forces of the Sun and Moon
and a complex combination of centrifugal force and continental buoyancy were involved. Simple calculations demonstrated that these forces were inadequate to move
continents. Because of a lack of an adequate force, the concept of mobile continents
lost support for a time.

Continental drift was revived after World War II in an entirely new guise with a new
source of energy. The new concept, called plate tectonics, suggested that not only did
the continents move, but that they are carried along on a mobile seafloor. Thus, the entire outer shell of Earth is in motion and the energy source comes from within Earth. This
energy is provided by thermal convection of heat from Earth's interior. The heat from
convection reaches the surface at the oceanic ridges where new seafloor is created by volcanic activity. The seafloor then spreads away from the ridges to be forced into the earth
at oceanic trenches, where the shell or plate of rock is reabsorbed.

Support for the concept of plate tectonics came from oceanographic studies and studies of Earth's magnetic field. However, acceptance of plate tectonics by the scientific community required confirmation from yet another independent source of evidence. The evidence was provided by the study of earthquakes at the oceanic ridges and trenches. The application of fault-plane solution techniques at these locations allowed a test of the motion of the seafloor. The fault-plane solution studies confirmed the plate tectonic hypothesis, showing clear evidence of movement of the seafloor away from oceanic ridges and toward seafloor trenches.

KEY WORDS

Atlantis	Holmes	Snider-Pelligrini
Bacon	land bridges	Sykes
continental drift	Matthews	Taylor
convection	oceanic ridges	Tonga trench
Curie temperature	paleomagnetism	Vine
dipole magnet	Pangaea	Wegener
Eotvos	polar reversal	Wilson
fault-plane solution	*polfluchkraft*	
Glossopteris fauna	seafloor spreading	

C H A P T E R   7

Journey to the Center of Earth

INTRODUCTION

Earth's interior has been shrouded in mystery for much of human existence because it is inaccessible. Most of our direct knowledge comes from the near surface, from exploration of caves or underground mining activities. But even here the deepest mines only take us down a mile or so beneath the surface where we still find solid rock. What is it like below this? Once again, the study of earthquakes provides some answers.

CAVES AND HOLLOW PLACES BELOW: IDEAS ABOUT EARTH'S INTERIOR

The idea that much of Earth's interior is hollow or open space is a very ancient one, which has persisted, at least in fiction, through the works of Jules Verne in the late nineteenth century. Hollow space can be inhabited, and perhaps there is a link here, for early humans were cave dwellers. It was thought by the Navajos of the American Southwest that they in fact originated from underground. Another slant on the hollow Earth concept was the idea that the open places below were the abode of the dead, connected with the afterlife. This can be seen in Dante's *Inferno* where evildoers are sent to the dark caverns below to be at Satan's whim. Two hundred and fifty years later Galesius believed that earthquakes were caused by the thrashing about of the dead in their agony.

Certainly the most recent and widely known account of open space underground is Jules Verne's *Journey to the Center of the Earth*. Verne's novel extended the concept of open space underground all the way to Earth's center where a large ocean was supposed to exist. At that time Verne had to combat evidence already in existence that would make such a world inside Earth quite unlikely. It was well known from mining experience that in deeper mines it is actually warmer than at Earth's surface. Added to this fact was the simple observation that volcanoes brought hot lavas to the surface from underground.

Verne in fact had the entrance to his underground world in the crater of an extinct volcano, a curious choice. One character in Verne's book, mindful of such facts, objected: "But is it not well known that heat increases one degree for every 70 feet you descend into Earth?" This objection was answered by the good professor with, "Don't be alarmed by the heat, my boy, neither you nor anybody else knows anything about the real state of Earth's interior. . . . Were any such heat to exist, the upper crust of Earth would be shattered."

Yet not only was temperature perceived to be a problem for hollow places inside Earth, but also pressure from the weight of overlying rock masses. It was well known that open spaces underground such as caves and mines were notoriously unstable and prone to collapse. Thus, the deeper one goes, the greater the weight of overlying rocks and the greater the tendency of open spaces to collapse.

Another widely held belief was that Earth's interior was liquid, given increasing temperatures and evidence from lavas brought to the surface. This idea is again expressed through one of Verne's characters: "All the matter which compose the globe are in a state of incandescence; even gold, platinum, and the hardest rocks are in a state of fusion."

Even by the end of the nineteenth century the widest held belief in scientific circles was that Earth's interior consisted of solid homogeneous rocks. This was certainly the assumption on which Lord Kelvin based his estimate of the age of Earth. Lord Kelvin calculated the planet's age based on the idea that it was a homogeneous solid sphere that had cooled from an originally molten state. It would take the study of earthquakes and earthquake waves to finally dispel the early myths and resolve the true nature of Earth's interior (Box 7.1).

SOLID AS A ROCK

A very compelling reason to support the concept of a solid interior of Earth is the behavior of earthquake waves. P-wave velocity varies according to the physical properties of the material it passes through. The lowest velocities are in gases, where we perceive P-waves as acoustic, or sound, waves. Higher velocities occur in liquids, and the highest velocities in solids. Since the earliest days of recording earthquake waves, seismologists have observed that P-wave velocity increases inside Earth, supporting a solid state.

Let us suppose an earthquake occurs beneath Earth's surface (Fig. 7.1). The P-wave is recorded at three different stations, A, B, and C. If the material in Earth were liquid, velocity should decrease after the waves pass through the solid crust. That is, it would take waves longer to reach stations B and C (B'C') than if the waves traveled through solid material with constant physical properties (B,C). What actually happens is that the waves speed up (B"C") arriving sooner than if they had traveled through solids with constant properties. This suggests that, far from being liquid, the material beneath the crust is not only solid but is even more rigid and dense than the solids forming the crustal rocks.

Another piece of crucial evidence exists to suggest that Earth's interior is not liquid beneath the crust. This is that S-waves generated in earthquakes are able to reach seismograph stations after traversing Earth's interior. It is known from theory and laboratory experiments that S-waves cannot travel through liquids. Hence, based on the behavior of earthquake waves traveling through Earth's interior, we have convincing evidence that Earth's interior is solid rock. But there is also evidence that although solid, Earth's interior is not homogeneous.

BOX 7.1

A Peek into Earth

The 1980s saw the beginning of ambitious programs in Europe and the United States to drill deep wells into Earth's crust. By the mid-1980s several super-deep wells were either in the planning stage or actively being drilled. The Kola well near Finland surpassed 11 kilometers in depth by 1985 and revealed a first-ever direct look at the conditions and rocks of the mid-crust. Surprises included mineral-rich deposits of nickel, copper, and lead at depths of 7 to 9 kilometers beneath the surface; higher than expected heat flow from crustal heat sources; and discovery of the true nature of crustal boundaries determined from seismology. The well was located on the Kola Peninsula of Russia east of Finland, penetrating rocks of the upper to mid-crust (A). To facilitate drilling during bitter winter conditions the work was done inside a special building complex (B). A core from the well (C) shows copper-nickel ore (Kozlovsky, 1987).

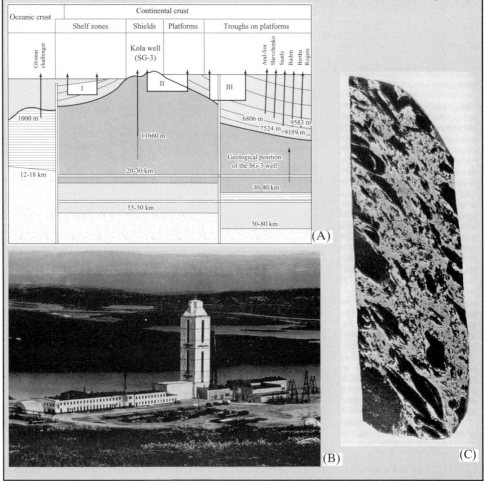

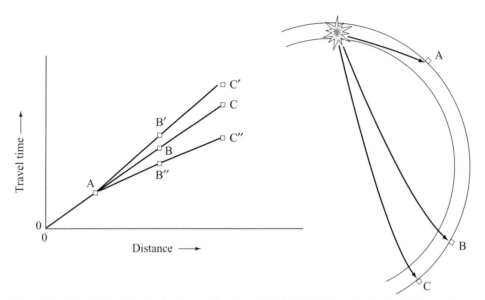

Figure 7.1   The effect of the physical properties of material in Earth on wave velocity. A decrease in velocity is represented by line *AB'C'*. Constant velocity is shown by *ABC* and increase in velocity by *AB"C"*.

A LAYERED EARTH: P-WAVE ECHOES

Within a quarter of a century after the development of operating seismographs it was becoming clear that Earth's interior was layered. This was the result of the observation of wave arrivals at stations that could only be explained by assuming that they occurred as a consequence of interaction with boundaries within Earth.

Bats cannot see their surroundings just as we cannot directly view Earth's interior. To adapt to a lack of light, bats have developed a sonar system. They rely on sound echoes to determine their distance from cave walls. Just as with sonar aboard ships, the bat sends out a high-pitched sound that bounces off cave walls. The longer it takes for an echo to reach the bat, the farther away the wall. The echo is the reflected sound wave. Seismic waves can bounce or reflect off layers inside Earth, and this is an important clue in determining the nature of Earth's interior.

The discovery of both reflected P- and S-waves meant that Earth's interior was not homogeneous. The existence of Earth's core had been discovered by 1906. The change of the material properties across the boundary between the mantle region and the core of Earth's interior provides a surface off which both P-waves and S-waves bounce or reflect (Fig. 7.2). Such P- and S-waves are designated by their path segments and the layers off which they bounce: PcP and ScS. The story gets even more complicated, for it is possible for part of the downgoing P-wave to bounce and convert to S-wave energy, i.e., PcS. This is because the P-wave compresses and extends material elastically in its wavepath (Fig. 7.3). As the P-wave strikes the boundary, it produces a component of motion in the reflected wavepath perpendicular to that wavepath—that is, the motion of an S-wave. If a P-wave approaches a boundary vertically its motion can only generate more P-waves when it encounters the surface. At other angles it is possible for P-

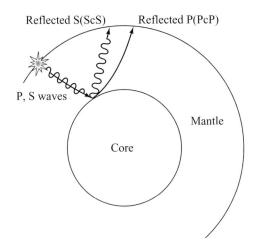

Reflected S(ScS)   Reflected P(PcP)

P, S waves

Mantle

Core

Figure 7.2   A section through Earth's interior showing the core–mantle boundary and the effect of the core boundary on earthquake waves. The waves reflecting off the core are indicated by lowercase *c*.

waves to convert some energy to produce S-waves and likewise for S-waves to produce P-waves. The multiplicity of waves that can be created at boundaries within Earth creates almost endless possibilities for wavepaths. It also increases the possiblity of detection of subtle boundaries within Earth.

A second phenomenon that occurs when a seismic wave encounters a boundary is termed *refraction*. This is the bending of the wavepath. An everyday example would be the case of a straw in a glass of water (Fig. 7.4). As the eye looks along the straw and into the water the straw appears bent. This, of course, is not the case, but is a result of the behavior of the light wave as it is affected by the change of material in its path. As light waves enter the water from the air, they are refracted or bent. This is because the physical properties of the air and water are different. Refraction of any type of wave occurs any time the density and, thus, velocity of two materials across a boundary are different.

The P-wave can be used as an example. Laboratory experiments have determined that P-wave velocities (Vp) are related to physical properties by the formula

$$Vp = \sqrt{(k + 4/3\mu)/\rho}$$

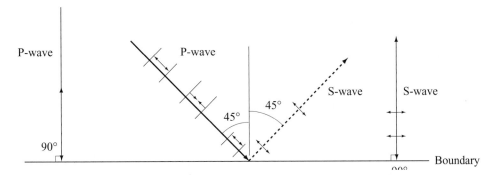

P-wave

P-wave

S-wave

S-wave

45°

45°

90°

Boundary

Figure 7.3   Reflected P- and S-wave behavior at a seismic boundary. A reflected P-wave may produce an S-wave with part of its energy as it reflects off a boundary.

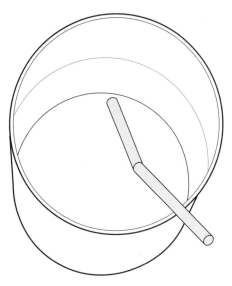

Figure 7.4   The optical illusion of the bent straw in a liquid. The apparent bending is the result of the refraction, or bending, of light waves between air and water.

As density (ρ) increases there is a corresponding and greater increase in μ (rigidity) and k (compressibility) as well. Thus, as ρ increases, Vp increases. Now imagine that there is a boundary between two materials (Fig. 7.5), with the material below the boundary having properties such that $\rho2>\rho1$, which also means that $Vp2>Vp1$. What happens to a seismic P-wave as it encounters the boundary? Consider a small segment of the wavefront encountering the boundary at an angle (Fig. 7.5). The part of the wavefront which reaches the boundary first (A in Fig. 7.5) speeds up as it crosses the boundary and rotates the whole wavefront, changing the angle of the wavepath, thus bending the

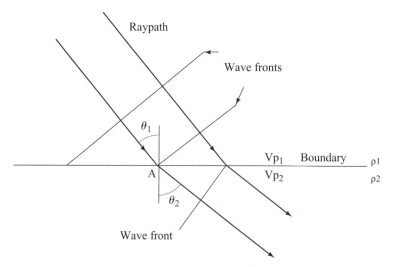

Figure 7.5   Refraction of a P-wave path at a boundary. θ_1 is the angle between wavepath and perpendicular to the boundary in the upper medium. V_{p1} is the P-wave velocity in upper medium; ρ_1 is density of upper medium.

direction of the wave. This is refraction and can be described by a relationship known as *Snell's law* between wavepath direction and velocity:

$$\frac{\sin \theta_1}{V_1} = \frac{\sin \theta_2}{V_2}$$

Although originally derived to explain the bending of light waves by Willebrord Snell, this relationship applies equally well to seismic waves. Snell's law is a powerful predictive tool, for it allows one to either predict changes in wavepaths, if velocities are known, or even more interestingly to predict physical properties of materials from the velocities. Just as was seen from reflection from boundaries, an important conclusion from Snell's law is that refracted waves cannot exist in a homogeneous Earth. This was made clear by the discovery of the boundary between the outer layer of Earth, the crust, and the region below, the mantle.

The boundary between the crust and mantle was named after its discoverer, Andreas Mohorovicic and is generally shortened to Moho. Mohorovicic was a scientist studying earthquakes in southeastern Europe and noticed that quakes from this region produced a curious feature on seismograms. This was the appearance of an extra phase or wave in the earthquake signature, in addition to the normal P- and S-wave arrivals (Fig. 7.6). Mohorovicic decided this was due to the interaction of the P-wave with a boundary beneath Earth's surface. This interaction produced two P-wave arrivals that had traveled different paths. One of the P-waves was a simple wave traveling a direct path through the upper layer (crust) to a seismograph station (Fig. 7.6). The second P-wave arrival traveled down and was refracted across the boundary, in such a way that it bent parallel to the boundary, traveling along beneath it. As it did so, it generated waves that traveled back up to the seismograph station. This refracted wave is termed a *critically refracted wave*. How does this happen? It can occur for a wave approaching the boundary at only one angle, termed the *critical angle*. This can be understood by considering Snell's law once again.

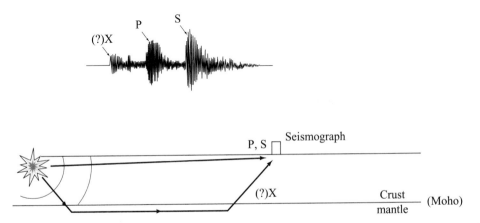

Figure 7.6   The creation of two P-waves traveling two different paths as a result of the existence of a boundary. This is termed *critical refraction*, and the X arrival is the critically refracted P-wave (Pn).

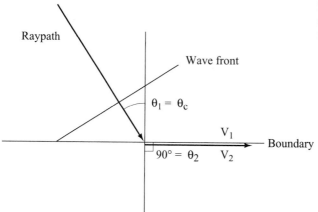

Figure 7.7   The critical angle is the sine of the angle and is equal to the ratio of the two velocities: V_1/V_2.

The first step will be to rearrange Snell's law into the form

$$\frac{\sin \theta_1}{\sin \theta_2} = \frac{V_1}{V_2}$$

This changes nothing, but is merely another convenient way of looking at the relationship. The next step is to investigate the critical angle, which is a special case of sin θ (Fig. 7.7). To do so it is necessary to transfer $\sin \theta_2$ to the right side of the equation so that it can be solved for $\sin \theta_1$. This can be done simply by multiplying both sides of the equation by $\sin \theta_2$. The result is

$$\sin \theta_1 = V_1/V_2 \sin \theta_2$$

But for the critical angle the result is $\theta_2 = 90°$, which is the same as $\sin \theta_2 = \sin 90°$ = 1. This simplifies Snell's law to one dealing only with the critical angle so that $\sin \theta_1 =$ sin θ critical = sin θc. So now the relation is

$$\sin \theta c = V_1/V_2$$

So for each pair of velocities above and below a boundary, there is one and only one unique critical angle at which a wave can approach a boundary and produce a critical wave traveling below and parallel to that boundary. This is a very important result, because it can be used to lead us to other information about Earth's interior.

STRUCTURE OF EARTH'S INTERIOR

The study of seismic waves has resulted in the discovery of three powerful tools useful in the study of Earth's interior: velocity, reflection, and refraction. Application of these tools has revealed a highly detailed picture of Earth's interior structure.

The change in velocity with depth reveals some rather complex structure. Velocity variation for both the P- and S-waves is well understood all the way to Earth's center (Fig. 7.8). Both sharp decreases and sudden increases in velocity are important. At depths of about 100 kilometers beneath the surface, the velocity of the S-wave drops, revealing the presence of a low velocity zone. This zone extends to depths of about 200 kilometers. The drop in velocity suggests a weaker and less rigid material, perhaps even in some small

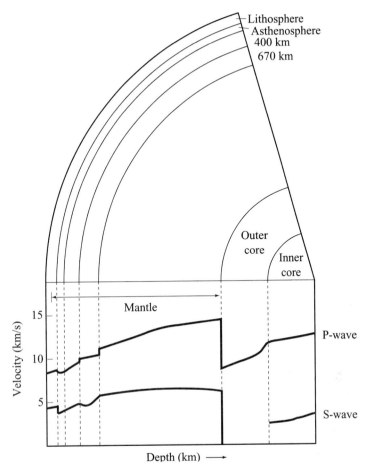

Figure 7.8   Velocity variation of P- and S-waves from the surface to the center of Earth. Changes in velocity reveal boundaries representing changes in density, composition, and phase (solid to liquid).

areas consisting of liquid (magma). This zone has since been termed the *asthenosphere* and is likely the weak zone upon which Earth's surface plates move.

Below the asthenosphere, the wave velocities increase again in a smooth fashion until about 400 kilometers when a sharp increase in velocity is observed. This is most likely due to a change in the physical properties of mantle rocks. The rock from the base of the crust downward is more homogeneous in bulk chemical composition, consisting almost exclusively of a silicate compound called *olivine*. At 400 kilometers the pressures are so extreme that it is thought that the crystalline structure of olivine collapses to produce a denser compound of the same chemical composition. The increased density results in an increase in seismic wave velocity. The change in density is recognized as a phase change and the depth at which it occurs is labeled the *400-kilometer discontinuity*. Another conversion occurs at 670 kilometers marked by yet a second velocity increase. At this depth the mineral likely collapses to form a final dense and stable compound with a structure like that of the high pressure silicate compound *perovskite*.

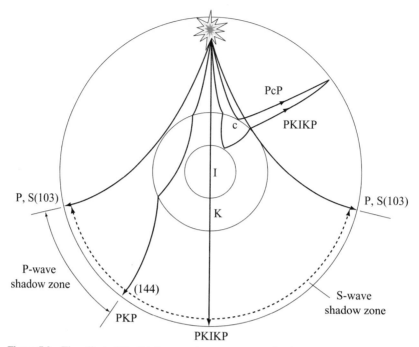

Figure 7.9 The effect of Earth's inner and outer core on seismic waves. Reflection at the core–mantle boundary (*c*) and inner core (*i*) is shown.

The lower mantle is quite uniform as indicated by a smooth increase in velocity until a depth of about 2900 to 3000 kilometers. At this depth there is a sharp drop in P-wave velocity, and the direct S-wave disappears, indicating the presence of an important boundary, that of the core. The disappearance of S-waves creates a shadow zone on the other side of Earth (Fig. 7.9). Stations at a surface distance of more than about 110 degrees away from an epicenter do not receive direct S-waves. This indicates that S-waves cannot traverse the core, suggesting the presence of either a liquid or gas. Because gases cannot exist under pressures present in the core region, at least part of the core must be liquid.

The P-waves reflect off the core boundary and return to the surface as the high-amplitude PcP wave (Fig. 7.9). P-waves that enter the core are refracted and travel across the core to reemerge on Earth's other side as the PKP phase. The bending of the wavepath leaves a gap at Earth's surface with no direct P-wave arrivals termed the *P-wave shadow zone*. This is analogous to the bending or focusing of light waves.

Earth's inner core can be detected by reflected or refracted waves. The P-wave reflected off the inner core is the PKiKP. Note that a small letter always indicates the surface of reflection, such as *c* for the outer core boundary and *i* for the inner core. P-waves refracted through the inner core are PKIKP. Such waves indicate that the inner core is solid, since they speed up crossing the inner core region, causing the wave to arrive earlier on the side of the Earth opposite the earthquake than they would have if the entire core was liquid.

Independent evidence supports a liquid outer core and solid inner core. The mantle is composed of silicate mineral compounds (e.g., olivine). Such compounds cannot exist as a liquid at the temperatures and pressures present at the mantle–core boundary. Yet the disappearance of S-waves clearly indicates a liquid state for the outer core. The only conclusion must be that the mantle–core boundary represents not only a phase change from solid to liquid but also a change in composition.

A study of the chemical abundance of elements in the solar system and Earth provides a useful clue to the composition of Earth's core. Considering the ten most common elements in the solar system, the element that is underrepresented in Earth's crust and mantle is iron. If the missing iron is concentrated in the core, it explains a number of observations. At the temperatures and pressures at the core–mantle boundary nonsilicate iron compounds would be liquid. Further support for an iron core for Earth is that by the time the inner core boundary is reached, the pressures and temperatures require conversion of the iron compounds to a solid state. The existence of a liquid outer core and a solid inner core required by results of study of seismic waves is then nicely supported by the presence of an iron-rich core.

The presence of an iron-rich core is supported by studies of Earth's density, which suggest that the core is composed of heavier elements and compounds than the mantle and crust. A core composed of a large amount of iron or iron compounds would explain the density data. Iron also fits well into theories about the role of the core in the generation of Earth's magnetic field. It is possible for a rotating iron-rich liquid core to generate a magnetic field. The magnetic field would not be fixed or stable because of the properties of a liquid, but would drift and change with time as does Earth's magnetic field.

X-RAYS INTO EARTH: THROUGH A GLASS DARKLY

A recent advance in the use of X-rays in medical diagnosis has been the CAT scan. CAT is short for computed axial tomography. The word tomography means picture (graph) and slice (tomo). A tomograph is a high-resolution, two-dimensional X-ray picture of part of the interior of the human body. Several adjacent slices are stacked to produce a three-dimensional image. Because of the large numbers of X-ray paths, computers are necessary to construct the images.

Seismic tomography of Earth's interior employs a very similar approach. Instead of X-ray paths, seismic wavepaths are used. Both body waves and surface waves may be used in solid-Earth tomographic studies. The variation in velocity as a response to physical properties of materials is the most frequently used aspect of the behavior of waves, for both surface and body wave tomography.

Surface wave velocity is strongly affected by wavelength. This is a result of the fact that surface waves respond to rock properties at depths that are proportional to wavelength. Thus, the longer the wavelength, the deeper the level to which surface wave energy will penetrate. Occasionally very large earthquakes will produce surface waves with wavelengths so long that the wave energy penetrates all the way to Earth's core. But such events are very rare, and so there is very little deep data available from surface waves. Surface wave tomographic studies are generally only applied to the crust–upper mantle region. Such studies complement tomographic studies using body waves for the lower mantle and core, for which a much larger body of data exists.

Seismic tomography depends upon changes along the wavepath. The earliest and most widely utilized path change was velocity. Velocity is affected by density and rigidity of rock materials such that warmer, weaker material will result in slower seismic waves traversing them, whereas colder, more rigid material will result in higher wave velocity. The problem is how to outline areas of unusual or anomalous velocity. One wavepath by itself may indicate the presence of a velocity anomaly, but in practice many wavepaths are required to determine the entire area of velocity change. The technique is to use as many wavepaths as possible. This requires many earthquake sources recorded at as many stations as possible to obtain the necessary density of wavepaths. Note that in Figure 7.10 that some raypaths will be higher velocity because they never cross the low-velocity body. In fact, rays F1–F4 crudely outline the anomalous body as they are the closest rays to the body that do not cross it. Rays S1 and S2, which cross the anomalous body, have correspondingly reduced velocities. Thus, the higher the density of stations and sources and thus possible raypaths that exist, the greater the resolution of the technique, provided the frequencies of the waves are high enough.

Seismic tomographic studies have been conducted at all scales and with varying degrees of resolution. Global studies have utilized large data sets such as that at the International Seismological Centre in Edinburgh, Scotland, in conjunction with high-speed computers. The results of global tomographic studies show clear support for the plate tectonic hypothesis. The images emerging from tomography show the existence of low ve-

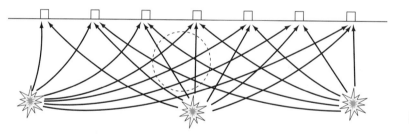

Anomalous body

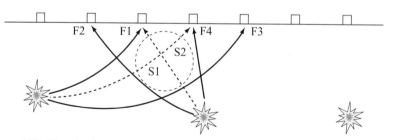

Figure 7.10   The seismic tomographic picture of an anomalous body of different density and thus velocity. The body is outlined by wavepaths of constant velocity (F_1–F_4), which do not cross the body as well as those that do and thus vary in velocity (S_1–S_2).

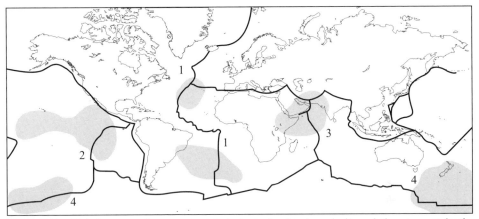

Figure 7.11   Global tomography supports plate tectonics by showing velocity variations to great depth within Earth. Low-velocity zones (lined patterns) exist beneath ocean ridges, suggesting high heat values. This map shows low-velocity zones at 250 kilometers below—(1) Mid-Atlantic ridge; (2) East-Pacific rise; (3) Carlsberg ridge; (4) East-Pacific rise–Southeast Indian ridge.

locities beneath oceanic ridges down to depths of about 250 kilometers (Fig. 7.11). High velocities generally correspond to colder rocks beneath continental interiors. The three-dimensional image of velocity anomalies suggests the presence of a convective flow pattern as the engine for plate tectonics: horizontal flow in the direction of plate motion, upward flow of warm material at the mid-ocean ridges, and downward flow at the oceanic trench sites of descending slabs of rock.

Seismic tomography can also image the downgoing lithospheric plates at subduction zones. A study conducted beneath Japan from the Japan trench to the Sea of Japan used P-wave velocity changes to outline the more rigid and dense lithospheric plate and separate its image from the surrounding less dense rock (Fig. 7.12). Negative values indicate higher velocities.

Seismic tomography has been applied at the regional scale to study volcanic hazards. The western United States is an area of moderate to high volcanic hazard. This has been brought home forcefully by the eruptions of Mount Lassen in California (1914 to 1917) and Mount Saint Helens (1980) in Washington State. Escaping steam and gases accompanied by tremors have raised alerts in California (Long Valley caldera), while elevated ground temperatures have caused concern at Mt. Rainier (Washington State).

Areas with signs of volcanic activity in the western United States have been studied by the federal government to determine whether magma exists beneath the surface, which could feed volcanic eruptions. An area with volcanic signs in California is Long Valley, not far south of Mono Craters. D. W. Steeples and H. M. Iyer with the United States Geological Survey used velocity measurements from seismic waves generated by distant earthquakes. These waves, as they passed through Long Valley, would also have to pass through any magma body present. The advantage of using waves from distant earthquakes is that the arriving waves would have to travel a nearly vertical path to reach the surface, thus giving a picture of velocity variations in a vertical section of Earth's crust (Fig. 7.13).

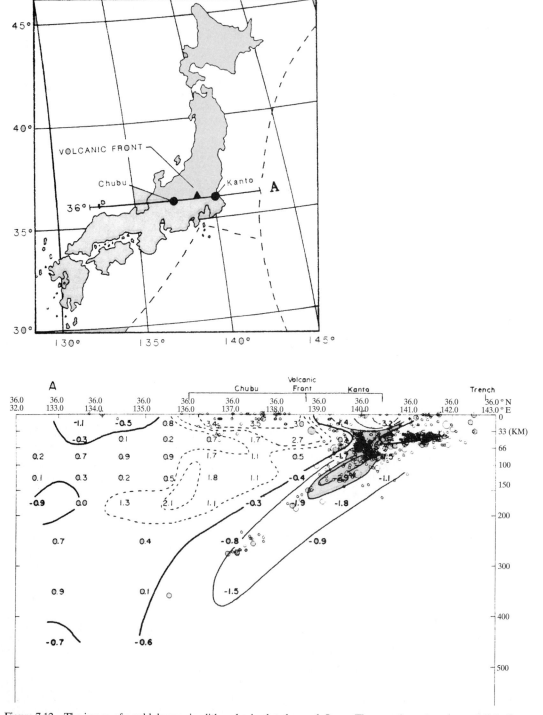

Figure 7.12 The image of a cold downgoing lithospheric plate beneath Japan. The negative values (e.g., –0.8) indicate high-velocity, colder material contained within the plate. Contour lines enclose anomalous velocity differences; circles are earthquake foci.

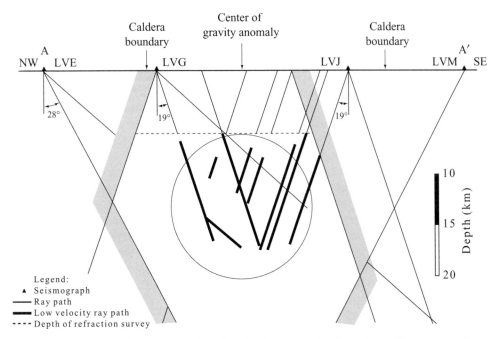

Figure 7.13   Results of seismic tomography at Long Valley caldera, California. Heavy lines on ray paths have lengths proportional to velocity delays along the path. An anomalous body with a density contrast of 0.18 grams per cubic centimeter and radius shown would enclose most of the anomalous low-velocity ray paths, and would account for the observed gravity field over Long Valley.

Results of the tomographic study of Long Valley agrees well with the results of density variation seen from measurements of changes in Earth's gravity field (Fig. 7.13). Both tomography and gravity indicate a region of low density, probably a liquid, at the same depths and with the same lateral extent. This region has a reduced P-wave velocity amounting to 15 percent less than normal, and a strong effect on S-waves. Thus, the Long Valley tomographic study lends support to the conclusion that a magma body exists at a depth of 5 kilometers beneath Earth's surface. Estimation of the size and depth of this magma body helps in determining the degree of volcanic hazard for this region.

SUMMARY

The nature of Earth's interior has long been a mystery. This has been so nearly up to the beginning of the twentieth century because of a lack of information. The study of earthquake waves has provided the necessary data with which to begin to understand fully the complex nature of Earth's interior.

The earliest studies of the interior relied on the more easily analyzed properties of waves: reflection, refraction, and velocity. Application of these properties to waves seen on seismograms resulted in the discovery of the major subdivisions of the interior: the crust, mantle, and inner and outer cores. The discovery of more subtle boundaries followed the development of seismic networks and other technological and analytical advances.

The development of high-speed computers in the 1960s and 1970s led to the use of wave velocity as a research tool in yet a different way. The application of computers allowed data analysis of thousands of wavepaths to map out areas of velocity change in Earth in both two dimensions and three dimensions. This approach to data analysis is termed *tomography* and has provided researchers with images of the interior analogous to X-ray pictures in medicine.

Seismic tomography, including the results of studies of both surface waves and body waves, has been applied to large-scale global research problems. The surface-to-core study of velocity changes has outlined a gross pattern that could represent the convective heat flow pattern of plate tectonics. Closer to the surface, images of subducted slabs have been revealed beneath oceanic trenches. Thus, tomography provides strong support for plate tectonics.

Seismic tomography has also been applied to the study of volcanic hazards. Tomographic imaging of the Long Valley caldera in California supports the existence of a rather large, shallow magma body, confirming the high level of volcanic hazard for that region.

The study of earthquake waves applied to the interior of Earth has been the primary tool in this area of research. It has provided crucial and supporting independent evidence used to discern the layered nature of Earth's interior, and the properties of those layers.

KEY WORDS

asthenosphere	Lord Kelvin	ScS
cat scan	Long Valley	shadow zone
caves	low-velocity zone	Snell's law
core	Mohorovicic	spinel
critical refraction	olivine	tomography
Dante	PcP	velocity
density	PKIKP	velocity anomaly
inner core	raypath	velocity discontinuity
Jules Verne	reflection	
iron	refraction	

PART III

EARTHQUAKES, EARTHQUAKE GEOGRAPHY, AND SAFETY

CHAPTER 8

Great Historic Earthquakes

INTRODUCTION

He who does not study the past is doomed to repeat its mistakes.

Anonymous

A study of the large and damaging earthquakes of the past is useful in that information learned can be applied to reduce casualties and damage in future events. Each time a damaging earthquake occurs new data are gathered by scientists, engineers, and others. These data are useful in improving the safety of buildings, planning where to build and not to build, and knowing how to prepare for an earthquake so personal safety can be improved.

The most useful data are from the effects caused by large damaging quakes, often called "great" because of their size and impact. These tremors have been with humans throughout history, but most of the early events are difficult to use or assess because of a lack of detailed information. An exception to this is an earthquake that occurred during the time of the late Roman Empire, about which much has been learned.

KOURION

Scientific expeditions to the island of Cyprus have uncovered the ruins of a Greco-Roman city that ceased to exist suddenly. Clearly, the city of Kourion had been witness to some kind of disaster, for the skeletons of people showed them huddling as if for protection. Furthermore, the blocks of buildings scattered about suggested wholesale collapse of structures. Evidence gathered from coins found at the site indicate that this catastrophe occurred sometime between late A.D. 364 and mid-367. This time frame matches the occurrence of a large earthquake in the eastern Mediterranean region reported by the Roman historian Ammianus Marcellinus. The earthquake occurred just after dawn of July 21, 365.

Evidence from the archaeological site and from the historical reports were utilized to reconstruct the earthquake. The degree of damage at Kourion allows application of the Modified Mercalli Intensity scale to estimate the size of this tremor. It seems clear that all building structures that have been excavated were totally destroyed. These were simple block structures such as the Temple of Apollo (Fig. 8.1) that depended on the weight of the structure and the mortar between blocks to hold it together. Under strong ground shaking in larger earthquakes such a structure comes unglued and tumbles apart,

Figure 8.1   Destruction of the Temple of Apollo at Kourion, Cyprus. The direction of collapse of the temple was used to determine direction of the incoming wave, and thus the direction to the epicenter.

like a pile of children's blocks. The widespread collapse of masonry structures at Kourion suggests an intensity of IX or X on the Modified Mercalli scale. This correlates with an earthquake of Richter magnitude of approximately 7.

The examination of the manner of collapse of the Temple of Apollo and other buildings in the area around Kourion was used by researchers to deduce the approximate location of the earthquake epicenter. The Temple of Apollo collapsed to the north and east, suggesting that the earthquake waves arrived from the southwest, the direction of the epicenter. Walls of buildings in Paphos, located to the west of Kourion, had fallen to the north and west (Fig. 8.2). Using these directions as vectors, their intersection to the south yields an approximate epicentral location. The epicenter lay 30 miles southwest of Kourion, beneath the floor of the Mediterranean Sea. This location coincides with an active seismic zone that is the boundary between the African and Anatolian tectonic plates. This oceanic location is further supported by historical reports of seismic sea waves associated with this earthquake.

Seismic sea waves are also called *tsunamis*, a term borrowed from the Japanese (Fig. 8.3). These waves will often occur in association with earthquakes that are at shallow depths beneath the seafloor. If a fault scarp forms and the seafloor is displaced upwards, the entire column of water above will be affected (Fig. 8.3). This uplift creates sea waves that travel outward at velocities that can exceed 600 miles per hour! A remarkable feature of tsunamis is that they have a very long wavelength and low amplitude at sea. The long wavelength is a consequence of the large area of ocean floor displaced during

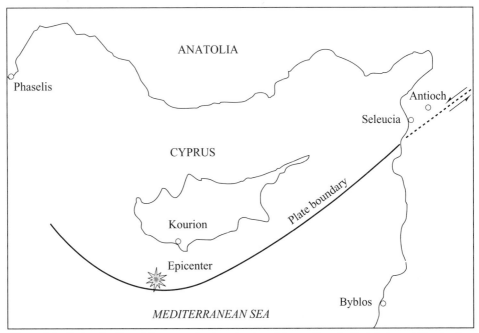

Figure 8.2   Determination of the approximate epicenter of the earthquake that destroyed Kourion in A.D. 365. The epicenter was close to a plate tectonic boundary, a site that can generate large and destructive quakes.

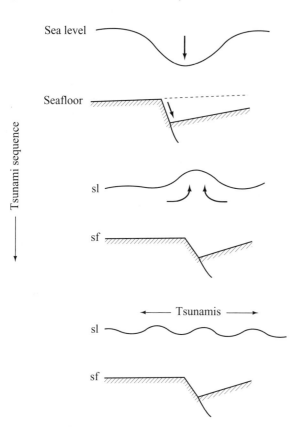

Figure 8.3   Tsunamis, or seismic sea waves, are produced by a fault that breaks the seafloor, displacing it and the column of water above it.

the earthquake. As with earthquake waves, the larger the source the longer the wavelength that can be generated. Ships at sea might encounter the low-amplitude tsunami waves without realizing it. As a tsunami approaches the shallower waters of a coastline, the drag with the seafloor slows down the waves and causes wave heights to increase. Waves as high as 100 feet can occur as they strike the coastline with tremendous force. The quake that struck Kourion created a tsunami that caused further horror at that unfortunate city and traveled across the Mediterranean to strike Greece and Egypt as well.

There is a lesson to be learned here: "Earthquakes don't kill people; buildings kill people." This is a saying dear to the hearts of structural engineers. While this is true, unfortunately it is not the only cause of loss of life in earthquakes, as can be seen from Kourion. Seismic sea waves are a serious threat for shoreline communities around the Pacific, Mediterranean, and elsewhere.

BASEL

The region of central and northern Europe is not very seismically active. Nevertheless, occasional smaller tremors are associated with the Rhine region. The Rhine River valley represents a downfaulted trough modified by surface erosion. This depression has an average north-south trend from Switzerland to the North Sea, bending to the northwest before it reaches the coastline. The movement on the trough-bounding faults has continued to the present with occasional large damaging events.

On October 18, 1356, at about ten o'clock at night, a large earthquake occurred somewhere in the vicinity of Basel, Switzerland, which lies on the Rhine River. The tremor quite likely occurred on one of the trough-bounding faults, and it shook Basel with a Mercalli intensity of XI, the same level reached at San Francisco in 1906. Extensive damage occurred from ground shaking over a large region around Basel. To the south in Berne, Switzerland, 40 miles from Basel, the arches of the great cathedral were cast down and part of the bell tower was lost (Fig. 8.4). Eighty castles within a radius of 25 miles (40 kilometers) were destroyed, and extensive damage occurred to structures in villages within the same radius and in Basel as well. At least 200 to 300 people died, and the ground shaking was followed by fire that continued for several days.

Lesson: You don't have to live in earthquake country (e.g., California) to be affected by damaging earthquakes. For this reason it is important to know the earthquake history of an area. The primary effect from the Basel tremor was ground shaking, which destroyed or damaged numerous buildings. Many shattered structures were finished off by fire. Earthquake-triggered fires have been a major problem over the centuries.

Figure 8.4   Both the bell tower of the cathedral at Berne, Switzerland, and the cathedral arches were destroyed in the earthquake of 1356.

SHANSI

The Hwang River of northern China has cut its way down into soft powderlike material called *loess*. This is windblown dust created by the grinding action of glaciers and then lifted by the wind and blown out across dry, cold plains. In the Hwang River region this material forms clifflike bluffs that are nevertheless soft and easily carved with simple tools. It is here that the Chinese made their homes and built their cities, carving numerous dwellings into the loess (Fig. 8.5).

On January 23, 1556, at about five in the morning, an earthquake estimated at magnitude 8 struck the region, collapsing the loess dwellings and causing landslides that buried survivors. The loss of life was carefully documented in Chinese imperial records and has been put at about 830,000. This is the greatest loss of life due to an earthquake in recorded history.

Lesson: This is the lesson of Kourion repeated with a twist. Instead of unreinforced block buildings we had people living in cave dwellings, which are even weaker structures. Sadly enough, this lesson was repeated in the same region in 1920. Somewhat farther to the west at Kansu (Fig. 8.5) another large tremor (Ms 8.5) again collapsed loess caves and this time took about 200,000 lives. Ironically, it is the fertility of the loess as well as the ease of carving it that continues to attract people to the region.

Figure 8.5 This map of China indicates locations of the great 1556 Shansi and 1920 Kansu earthquakes.

JAMAICA

Most people when they visualize Jamaica picture the sun-kissed beaches of a vacation resort. Although accurate, few would believe that this island is in earthquake country. Jamaica sits adjacent to a plate boundary between the North American and Caribbean tectonic plates. Movement of these plates generates occasional large and damaging tremors, as was the case in 1692.

Port Royal, Jamaica, was founded because of its geographic position. The port and protected harbor are formed by a long arm of sand stretching out from the shoreline, enclosing a bay (Fig. 8.6). This sand-bar structure was at one and the same time a blessing and a curse. The fact that it existed created a premier harbor, important to the English government. But the material from which it was made would prove to be a great liability. Even from biblical times there existed a sense that sandy material did not make the best of foundations. Jesus told Peter: "You are the rock; upon you I build my church." In other words, stable foundation material is crucial. This sense of solid foundation had been borne out by experience. The sand beneath Port Royal was loose and water saturated and had always been a problem. Block buildings erected on such a foundation have a tendency to settle and separate along the mortar between blocks, creating cracks. A worse surprise was in store for the inhabitants of Port Royal, a phenomenon known as *liquefaction*.

Liquefaction is a term that describes what happens to loose water-saturated sand when it is vigorously shaken. The sand particles lose contact with one another and float in the water, giving the entire mixture the property of water. Anything originally supported at the surface of the ground such as buildings, signs, wagons, or even people either float or sink. The ground shaking creates excess pore water pressure that may also force sand and water to the surface to form spectacular fountains that build up small volcanic-like cones of mud and sand around the vents. Such was the potential for disaster at Port Royal, Jamaica.

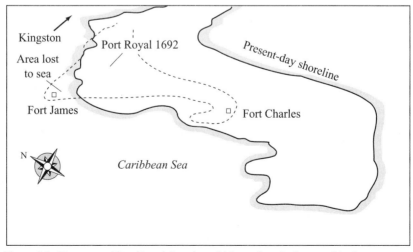

Figure 8.6   The town of Port Royal, Jamaica, which was largely destroyed in the earthquake of 1692.

On June 7, 1692, just before noon, a devastating earthquake shook Port Royal and the entire island of Jamaica. The magnitude has been estimated at about 8 because of the duration and severity of ground shaking, and the length of time over which aftershocks persisted. Three months after the tremor there were still three aftershocks a day. This is typical of large and damaging quakes.

Liquefaction was widespread throughout Port Royal, and buildings that were not damaged by ground shaking were swallowed up by liquefaction. Up to one-third of Port Royal simply vanished, sinking beneath the waters of the bay (Fig. 8.7). The vertical movement of this part of Port Royal was due to yet another aspect of liquefaction called *lateral spreading*.

Liquefied sands, under the force of gravity, cause high areas to sink and spread and for low areas to rise. Thus, in 1692, part of the sand bar beneath Port Royal sank below the water until it was about 12 feet beneath the water surface. At the same time part of the harbor floor rose, making the harbor shallower. For many years thereafter, this part of Port Royal—streets, houses and all—could be seen beneath the water by boat passengers crossing the harbor.

Lesson: Approximately 2500 people died in the 1692 Jamaica earthquake, many of them clearly as a result of burial by liquefaction (Box 8.1). Many areas where people are building today such as along beaches and in river valleys are liquefaction-prone areas.

LISBON

> The grinding of the walls, the fall of churches, the lamentable cries of the inhabitants, join'd to a perfect darkness occasione'd by the dust, made one of the dreadfullest scenes of nature.
>
> English merchant, and earthquake survivor

Figure 8.7   Up to one-third of Port Royal disappeared during the 1692 tremor because of liquefaction of the sand spit on which the town was built.

BOX 8.1

Eyewitness to Liquefaction

Not only buildings but also people were literally swallowed up by liquefied sands at Port Royal, Jamaica, as a result of the 1692 earthquake. This was dramatically captured in the words of eyewitnesses:

> [W]hole streets with inhabitants were swallowed up by the opening earth, which then shutting upon them, squeezed the people to death. And in that manner several are left buried with their heads aboveground.

> Some were swallowed quite down, and cast up again by great quantities of water; others went down and were never more seen.

Somewhere to the southwest of Lisbon, Portugal, on November 1, 1755, under the waters of the Atlantic Ocean, Earth's crust broke and slipped, releasing energy that has been estimated to be equivalent in magnitude to Ms≥8. The resulting ground movements caused by seismic waves were felt over virtually all of Europe and North Africa. The potential area over which the earthquake could have been felt (including ocean) is estimated at 1,600,000 square kilometers. The tremor was felt as far away as Switzerland and Scotland, but the worst damage outside of Lisbon occurred in North Africa, where Algiers was almost totally destroyed. Tsunamis were generated that were most

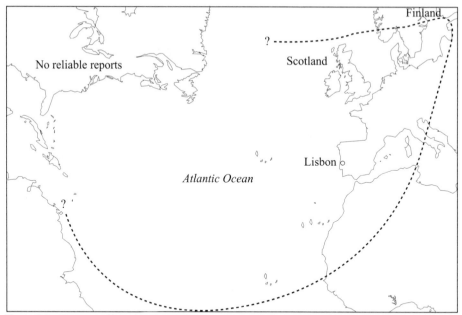

Figure 8.8 The area affected by the 1755 Lisbon tremor is indicated within the dashed line. Effects include ground shaking, tsunamis, and seiches.

destructive in Europe, but they traveled across the Atlantic to lap up on the shores of the West Indies (Fig. 8.8). At Kinsale on the Irish coast a large surge of water swept away everything in its reach, while at Antigua Island in the West Indies and over 3000 miles (4800 kilometers) from Lisbon, a sea wave of 12 feet (4 meters) surged over the shore. Water in lakes rocked back and forth like water in a bathtub, a phenomenon known as *seiche*. Seiches were noted over much of Europe as far away as Finland and Scotland. All told it is estimated that at least 70,000 people lost their lives in the most disastrous earthquake to strike Europe in recorded history (Fig. 8.9).

The timing of the earthquake was perhaps most ironic as well as unfortunate. It occurred on a bright Sunday morning. November 1, 1755, was All Saints Day on the Church calendar, and in a very pious age Lisbon's churches were crowded when the first shock waves arrived at 9:40. According to survivors, the priests had begun the *Gaudeamus omnes in Deo* when the walls began to rock. Although some fled into the streets, few people escaped. Within moments a stronger shock arrived and continued for perhaps several minutes. The churches and large public buildings of unreinforced block collapsed, killing those within and showering survivors in the streets with debris. Dust clouds from the fallen masonry clogged the air.

Many of those still alive gathered along the waterfront for safety or in open spaces away from buildings, for the ground continued to shake intermittently. At about 10 o'clock the sea withdrew, and then came rushing back in with the first of three great tsunamis. The waves engulfed the waterfront, causing hundreds of casualties and obliterating everything within the waves' path.

Shortly after the first shock, fires from church candles and from home cookstoves began to spread out across the city. Fire engulfed houses, churches, and palaces, sweep-

Figure 8.9   A print of the destruction of Lisbon, Portugal by the great November 1, 1755, earthquake and resulting fires and tsunamis.

ing away priceless treasures and leveling what the tremors and tsunamis had left. So devastating was the earthquake and its aftereffects that the buildings of the city of Lisbon today date almost entirely from after 1755.

Lesson: Be careful where you build. The tremor of November 1, 1755, was not the first affecting Lisbon. A previous earthquake in 1531 was reputed to have destroyed thousands of homes and all the churches in the city. This earlier earthquake was also accompanied by a seismic sea wave.

NEW MADRID

The earthquakes that struck the Mississippi River valley in 1811 and 1812 were perhaps mirrored in the Old Testament: "Every valley shall be exalted, and every mountain and hill shall be made low; and the crooked shall be made straight, and the rough places plain" (Isaiah: 40:4). Although taken out of context, this passage sounds like a description of the sort of deformation that occurs in the very largest earthquakes. Such was the case in the sequence of tremors named after New Madrid, Missouri.

The Mississippi River valley lies far away from any plate boundaries today, well within the North American tectonic plate, and thus in an area where large damaging earthquakes are generally infrequent to absent. Yet well below the soft muds and sands of the river valley lies a fault system that represents an aborted attempt to form a plate boundary in the distant geologic past. It was a fault in this zone that ruptured at 2 A.M. on December 16, 1811. Because these tremors occurred well before the establishment of seismograph stations, an estimation of magnitude is not clear-cut, but based on the ground-shaking effects and resulting damage, these tremors were unquestionably Ms≥8.0.

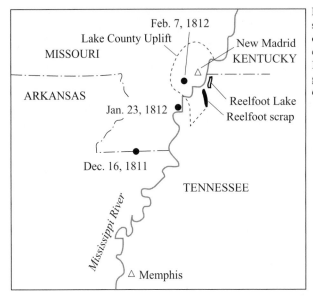

Figure 8.10 A map of the region of strong effects (meizoseismal zone) of the 1811–1812 New Madrid earthquakes. Lake Saint Francis and Reelfoot Lake were created by ground movement and changes in elevation caused by the quakes.

Rather remarkable vertical movements accompanied these tremors, which resulted in permanent changes in the landscape. A large area around the Mississippi channel and on both sides subsided, forming swampy areas and lakes. The ground-surface elevation was permanently lowered as much as 15 feet (5 meters) and created topography that favored the formation of Lake St. Francis west of the Mississippi and Reelfoot Lake to the east (Fig. 8.10). Smaller regions to the west of the depression rose as much as 20 feet (about 7 meters). Waterfalls were temporarily created in the bed of the Mississippi River, and the caving of soft banks and trees into the river created tremendous turbulence. Liquefaction and accompanying sand boils or mud volcanoes were widespread. Because of changes in the river channel of the Mississippi, the town of New Madrid largely disappeared. It must have seemed like the end of the world.

The December 16 shock was strong enough to keep people outdoors despite the cold weather. Aftershocks continued frequently until 7 A.M. when another large tremor was felt, as strong as the first. A third large quake occurred at 11 A.M. and may have been the strongest yet, or perhaps the closest to New Madrid. The sequence of aftershocks continued until January 23, 1812, when at 9 A.M. another strong shock apparently equal in size to the 2 A.M. December 16 event occurred. The last large earthquake in the series took place on February 7, 1812, at about 3 A.M. The larger shocks in the sequence were felt throughout the populated regions of the eastern United States and Canada and by at least one tribe of Native Americans on the present-day Nebraska-Kansas border. All together, it was estimated that the largest tremors were felt over an area of 1,000,000 square miles.

The December 16 quake knocked down chimneys in Cincinnati about 400 miles from the epicentral region. The tremor was also felt strongly in Washington, D.C., rattling doors and windows some 800 miles from New Madrid. The February 7, 1812, shock caused even more damage at Cincinnati. At Louisville, Kentucky, damage was widespread with toppled chimneys and parapets.

Lesson: The earthquake threat is clear along plate tectonic boundaries such as in California, yet areas far from such boundaries like the Mississippi River valley may also be subject to less frequent but still damaging tremors. The source of the Mississippi tremors lies in ancient fault systems that allow stress readjustments along a failed tectonic plate boundary, now buried deeply by river sediments and younger rock materials. Knowledge of the existence of both previous earthquakes and of earthquake geography is important in understanding earthquake hazards.

SONORA, MEXICO

On May 3, 1887, a large earthquake occurred just south of the Arizona border in the Mexican state of Sonora. The tremor was accompanied by the formation of a fault scarp at least 50 miles (80 kilometers) long. The magnitude estimate based on the length of the fault scarp is Ms 7.4, which makes the tremor one of the largest in this region in historic times.

Ground shaking from the Sonora tremor was felt as far north as Santa Fe, New Mexico, as well as in Phoenix and Yuma, Arizona, and in a southerly direction all the way to Mexico City, a distance of 700 miles (1120 kilometers) from the epicenter. Damaging ground shaking was confined largely to Sonora and southern Arizona (Fig. 8.11). A total of 51 deaths resulted from the tremor, all in Sonora, and the majority of building collapse was confined to the epicentral area as well. Forty-two of the fatalities were in Bavispe, Sonora, and most probably occurred when people rushed into the colonial church just before the roof collapsed (Fig. 8.12). This structure was of unreinforced masonry and showed damage typical to that of early colonial churches in both Sonora and Arizona. Significant damage also occurred to the church structures at San Xavier del Bac in southern Arizona. Collapse of unreinforced block or adobe structures was widespread throughout the region. The town of Charleston, Arizona, was totally abandoned, as not a single building was habitable after the tremor.

Serious secondary effects were noted throughout much of the high-intensity ground-shaking region, some of which were especially close to Tucson, Arizona. Liquefaction occurred in the San Pedro valley, 30 miles southeast of Tucson, while extensive rockfalls and landslides were noted in the Santa Catalina Mountains, which lie on the eastern edge of the metropolitan area of Tucson today. Significant rockfalls were also observed on South Mountain just to the south of Phoenix.

Lesson: Once again a damaging earthquake occurred in an area not noted for tremors. Study of the region of the 1887 fault scarp has revealed a second, older scarp. Thus, this was not the only large tremor to have hit the region. The challenge will be to try to use this information to understand clearly the hazard in such locales. The results of a future event could be catastrophic, for Tucson is the second largest city in Arizona.

SAN FRANCISCO

I was awakened at 5:13 on the bright sunny morning of April 18th . . .; the wave which woke me was gentle enough, but the next one, like the bump of an express train, seemed a little severe . . .; the bedroom on the second floor swayed like a ship in a hurricane. A lantern standing in the hall leaped in

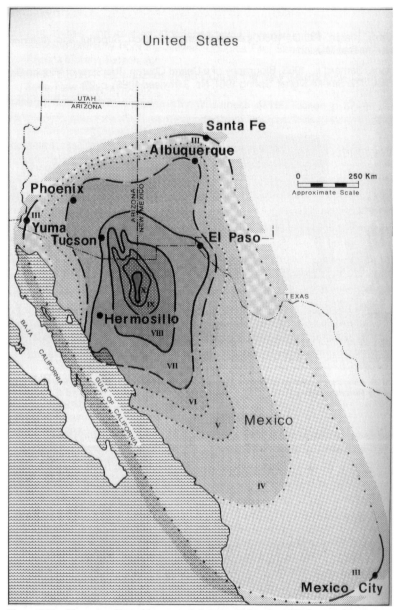

Figure 8.11 The area affected by ground shaking during the 1887 Sonora, Mexico, earthquake. Landslides were triggered by ground movement as far north as Phoenix, Arizona.

Figure 8.12   A photograph of the remains of the Bavispe, Sonora, church destroyed by the 1887 Mexican tremor.

through the open door. Pictures swayed, earthenware leaped about. Some mighty force seemed to hold the house, and to be trying to whip the ground with it. . . . The older boy, who was sleeping on the roof, clung on as to a runaway horse. As things became a little calmer, he shouted down: "The church is falling! The gymnasium is caving in! . . ." Then I knew that we had had an epoch-making earthquake.

David Starr Jordan

The earthquake just described, which became known as *the* San Francisco earthquake, was centered on the San Andreas fault and, as the eyewitness account implies, leveled much of the city of San Francisco in 1906. The ground shaking was actually timed at 40 seconds and was long enough to cause tremendous damage over a wide area. Areas hardest hit by ground shaking were those along the shoreline where buildings were erected on soft bay muds. Even more serious, however, was the damage to the city water supply system. City water had been brought in from reservoirs to the south by three pipelines. One of the three lines ran directly along the San Andreas fault for a distance of 7 miles and was essentially destroyed. The other two pipelines were broken at points where they crossed marshes. Water pipes in the city were destroyed where they crossed fill materials; the net result was that when the ground shaking ceased, the city of San Francisco was without a water supply. This was to have dire consequences, for very soon after the ground shaking had abated, several fires began to spread.

Fire officials recognized the problem, and in less than an hour after the quake struck they asked the U.S. Army stationed at the Presidio in northern San Francisco for dynamite to add to the supply from the city's engineering department. The strategy was to dynamite buildings and create fire lanes to keep the flames from spreading. As events would later prove, however, dynamiting was not enough. The situation was described as follows:

All night the city burned with a copper glow, and all night the dynamite of the fire fighters boomed at slow intervals, the pulse of a great city in its agony.

James Hopper, *The San Francisco Earthquake and Fire*

Figure 8.13   The fires that followed the 1906 San Francisco earthquake destroyed much of the city.

The fire was to continue to burn for three more days, lifting clouds of smoke into the sky that could be seen for many miles. All in all, over 4 square miles and more than 28,000 buildings in downtown San Francisco were destroyed (Fig. 8.13). This made determination of casualties difficult. Estimates ranged from about 700 to more recent figures of between 2500 to 3000. These numbers reflect casualties outside of San Francisco as well. It is unquestionably the greatest loss of life from an earthquake in the United States.

Lesson: Fire is frequently an unfortunate consequence of earthquake ground shaking. Precautions have been taken since 1906 to ensure that the San Francisco water supply is not so vulnerable. As can be seen from more recent tremors such as Loma Prieta, this does not eliminate the threat of fire from sources like broken gas lines.

San Francisco will continue to be at risk from future earthquakes along the San Andreas fault. Part of the city lies on top of the fault, and most of the rest is well within the reach of severe ground shaking.

TOKYO

A great earthquake struck Tokyo just before noon on September 1, 1923. Estimated at magnitude 8.2, the epicenter occurred beneath Sagami Bay (Fig. 8.14), just 50 miles southwest of the city. This resulted in what is thought to be the greatest loss of life in a natural disaster in Japan.

Some 143,000 people died as a result of the ground shaking, a tsunami, and subsequent fires. The fires were even worse than in San Francisco, with an estimated 134 fires in Tokyo resulting from the tremor, most starting as a result of overturned cookstoves that were preparing the noon meal. Because many of the homes were of wood and had been destroyed by the ground shaking, the wood from the houses ignited like piles of kin-

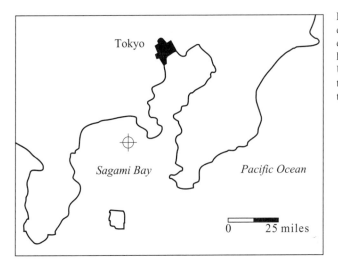

Figure 8.14   A map of the epicentral area of the 1923 Tokyo earthquake. The epicenter was located beneath Sagami Bay. Uplift of the bay floor resulted in the generation of a destructive tsunami.

Tokyo

Sagami Bay

Pacific Ocean

0   25 miles

dling. The fires were out of control within minutes, and Tokyo, like San Francisco in 1906, had lost its water supply. The fires merged, created their own updrafts, and filled the sky with flames. The huge conflagration became a firestorm, an entity so intense that it sucked everything combustible into an enormous funnel of flame (Fig. 8.15).

Those who had survived the earthquake were driven out in front of the fires, seeking open spaces. As a result, a large number of people were forced into an open area of 10 acres, formerly occupied by a military clothing depot. The firestorm surrounded them, and in an instant the flames engulfed an estimated 40,000 people, leaving only 30 survivors.

To add to the misery and suffering caused by the tremor and fire, a 36-foot-high tsunami struck the coast of Sagami Bay. The wave swept away homes, railroad tracks, bridges, and roads. Hydrographic and geodetic surveys completed after the earthquake showed large relative movements of the land surface to the south, and striking changes in the water depth under Sagami Bay.

Lesson: Much was learned about earthquake-resistant design for buildings from the Tokyo earthquake. The Imperial Hotel, designed by Frank Lloyd Wright, had been built to resist earthquake damage and, in fact, rode out the tremor well. Fire was a major player, and again, as at San Francisco, a lack of reliable water supply created a desperate situation.

CHILE

Chile lies over the plate subduction zone between the South American plate and the Nazca plate. The surface expression of this boundary is the submarine Peru–Chile trench stretching north-south, parallel to the coastline (Fig. 8.16). Pressure had been building up along the boundary between the two plates by their inexorable movements. On Saturday morning, May 21, 1960, at 6:02, the surface between the plates began to break loose, resulting in the largest release of earthquake energy in the twentieth century. Subsequent calculations assigned a moment magnitude of 9.5 to the main shock. So large was

Figure 8.15   The greatest loss of life in an earthquake-caused fire was recorded in the 1923 Tokyo earthquake firestorm, where 40,000 people died.

the amount of energy released that the ground responded as though it were a rubber ball being squeezed by a large hand, first compressing, then slowly expanding, vibrating in this fashion all of the way to Earth's core.

The response along the trench was a lot more immediate and devastating. The 6 A.M. shock was actually a foreshock that began a sequence of tremors, including the mainshock and aftershocks, which lasted for days, extending fault slip over a length of 1000 kilometers (625 miles). The area that moved beneath the surface was about the same size as the state of California. Large areas of Earth's surface were permanently elevated and depressed parallel to the coastline (Fig. 8.17). An estimated 17,000 square miles were affected by vertical changes in elevation of as much as 13 feet up and 7 feet down.

The secondary effects of the quake sequence were profound as well. Landslides were numerous and some quite large. The river valley above Valdivia had been blocked by a mountain of earth 175 feet high, which threatened to unleash the water trapped behind it on the unfortunate populace below. Using earth-moving equipment, it took days to cut this barrier down and create a channel so that the threat was eliminated.

Another danger was created by movement of the seafloor. One of the most destructive tsunamis in many years spread out across the Pacific Ocean, traveling at an average speed of 442 miles per hour (Fig. 8.18). At about 15 hours after it left the Chilean coast it struck Hilo, Hawaii, killing 61 people, and then it continued on to Japan, arriving almost 23 hours later. Crashing ashore in the Japanese islands, the waves swept boats inland, battered the coastal villages, and killed another 185 people.

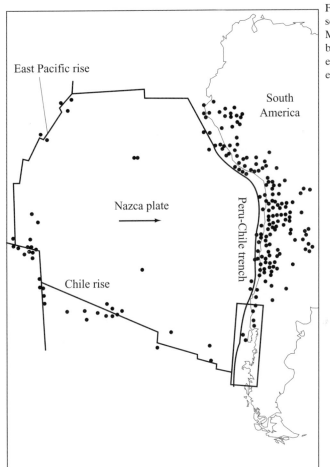

Figure 8.16 The plate tectonic setting of the Chilean coastline. Movement of plates along this boundary releases abundant earthquake energy. Each dot is an epicenter.

An unusual aspect of the Chilean tremors was that the release of earthquake energy seems to have been responsible for triggering a volcanic eruption. Two days after the main shock, Puyehue volcano erupted from a 1000-foot-long fissure. Although there are many cases of a volcanic eruption causing the ground to quake, this appears to be the first documented case of an earthquake activating a volcano.

Lesson: A interesting fact about the Chilean earthquake sequence is that the largest earthquake (Mw9.5) of the twentieth centruy did not result in the greatest death toll. Although between 2000 and 3000 people died, casualties could have been much higher except for circumstances. A large foreshock, which occurred 30 minutes before the main shock, had brought people out of their homes, which then collapsed during the main shock. Casualties in earthquakes are strongly dependent on when the tremor strikes. The November 1, 1755, Lisbon earthquake occurred at mid-day during a religious holiday, and these factors were certainly partly responsible for the large number of casualties that resulted.

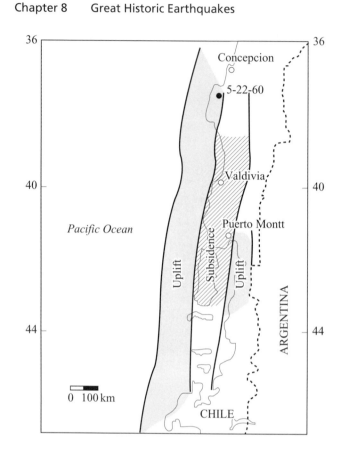

Figure 8.17   A map of permanent vertical changes in the land surface as a result of the 1960 Chilean earthquake. Location of 5-22-60 epicenter indicated.

ALASKA

The great Alaska earthquake occurred on Friday, March 27, 1964, and has ever since been known as the Good Friday earthquake as this was the Friday before the Easter holiday. At 5:36 P.M. the Pacific plate boundary beneath Prince William Sound slipped. The energy released was enormous, resulting in the largest earthquake to strike North America in the twentieth century, and exceeded in the world since 1900 only by the 1960 Chilean tremor of Mw9.5. Sound waves from the quake were heard as far away as San

Figure 8.18   The first floor of this home was swept away as a result of the tsunami generated by the 1960 Chilean earthquake.

Diego, and much of the North American continent rose several inches in response to the passage of earthquake waves. Just as in the Chilean quake, Earth down to its core vibrated like a struck bell.

Vertical changes in Earth's surface elevation that accompanied the Alaska earthquake were the largest in recorded history. A total of 48,000 square miles of seafloor and land dropped several feet, while another 60,000 square miles of Earth's surface rose (Fig. 8.19). A large area of the south coast of Alaska suffered severe damage. The hardest-hit communities were Valdez, Seward, Kodiak, and Anchorage.

The port city of Valdez was literally destroyed. As ground shaking began, the waterfront slid into the bay, carrying the pier with it and creating a giant wave, or seiche. The freighter *Chena* was docked at the time and, in response to the waves, the ship had a wild ride, rising as much as 30 feet, then dropping like a runaway elevator. Somehow the *Chena* survived. But Valdez was pummeled by the waves, which did extensive damage, even ripping the channel lighthouse from its foundation. But this wasn't all. Thirty minutes after the earthquake, the first tsunami wave arrived. Afterwards, more than half the buildings in Valdez were unusable or totally destroyed, and 32 people lay dead.

Seward, a major oil terminal and freight yard for the Alaska railroad, lies 125 miles (200 kilometers) southwest of Valdez. It was also devastated by ground shaking and tsunamis; in addition, because of the presence of oil tanks, large fires broke out as well. As part of the waterfront slid into the bay, oil pipelines ruptured and the contents ignited, causing nearby oil storage tanks to explode in flames. The combination of fire, ground shaking, landsliding of the waterfront into the bay, and tsunamis totally destroyed the Alaska railroad yards, overturning 125-ton locomotives in the process.

Figure 8.19   Vertical changes in the land surface due to the 1964 Alaskan earthquake.

Kodiak Island, 200 miles southwest of Seward, had little damage from ground shaking. However, the tsunamis engulfed and destroyed half the fishing fleet in port and two seafood canneries. Rather remarkably, only eight people died.

Northeast of Kodiak and close to the epicentral area is Anchorage. Anchorage was then and still is Alaska's largest city. An important seaport, the city lies on Cook Inlet at the head of a shipping channel. Ground shaking was severe, lasting about 30 to 60 seconds. Small frame houses survived the ground shaking well, but several large structures were destroyed, including the concrete control tower at Anchorage Airport and a brand new six-story apartment house, both of which collapsed. Downtown Anchorage suffered similar selective devastation. The new multistory J.C. Penney department store lost its facade, which consisted of large suspended concrete slabs. The slabs came cascading down into the street, crushing automobiles (Fig. 8.20). Although much of the rest of the business district escaped total destruction, a quarter-mile section of Fourth Avenue sank more than 10 feet, carrying shops and cars with it (Fig. 8.21). Perhaps the most stunning effect of the quake occurred in southwest Anchorage where part of a housing development, Turnagin Heights, slid laterally toward Cook Inlet. Seventy homes were destroyed.

Lesson: Where you build does matter. Much of the damage and loss of life in the Alaskan quake was the direct result of building on poor foundation materials. The underwater landslide that took the waterfront and pier at Valdez is only one example, whereas the Turnagin Heights landslide at Anchorage was another. For many years following the 1964 tremor, Turnagin Heights was a park area, and development was for-

Figure 8.20 Damage to the J.C. Penney store in Anchorage from the 1964 Alaskan earthquake.

Figure 8.21   Destruction of Fourth Avenue in Anchorage as a result of the 1964 Alaskan quake. A long stretch of the street sank more than 10 feet as a result of ground shaking.

bidden. Thus, the lesson in this case was clear: The best prevention is to be careful where you build.

PERU

On May 31, 1970, in the Andes, history repeated itself. The Andes lie along the Pacific plate boundary and are in fact due to movements of the Pacific plate as it is forced beneath South America. The towering peaks of the Andes volcanoes provide unstable slopes over which landslides travel down to the valleys far below. Such was the situation on May 31 when a magnitude-7.7 earthquake not far offshore shook the mountain slopes, breaking loose 100 million cubic meters of granite, ice, and sediments that cascaded rapidly downslope at velocities estimated at 175 to 210 miles per hour. On the lower slopes the landslide sought out and followed the valleys down the side of the mountain, splitting around and climbing over a 750-foot-high hill. Unfortunately, the villages of Yungay and Ranrahirca lay in the path of the roaring mountainous material. Survivors tell of an enormous mass of surging debris about 250 feet high, accompanied by a shock wave of compressed air. Nearly 30,000 people died, buried under the debris. The only survivors in the town of Yungay were a few residents who took refuge atop a ridge that was also the site of the town cemetery.

Lesson: Avalanches and landslides are not uncommon in the steep-sided valleys of the Andes. Nor was this the first such event in this valley. In 1962 a similar mass-movement covered Ranrahirca, resulting in 4000 deaths. The future will likely see more such disasters, for people will rebuild and continue to live in the valleys. Because people will still reside in hazard-prone regions, the strategy must be to learn how to cope with events such as earthquakes and landslides to protect lives.

MEXICO CITY

One of the most terrifying places to be in an earthquake is in a high-rise building. This is because the upper part of these structures can move as much as several feet back and forth, toppling furniture and filing cabinets. Nowhere was this better brought home than in Mexico City on September 19, 1985. Two hundred and twenty miles to the west off the Pacific coast of Mexico, the rocks broke and slipped, releasing the energy of a magnitude-8.1 earthquake. Within minutes the waves released by the tremor had reached Mexico City, and the ground began to shake.

Mexico City is the most populous urban area in the world, with approximately 20 million people. So it was fortunate that the quake occurred at 7:19 A.M., before most of the downtown office buildings were filled. The ground shaking in Mexico City lasted about 3 minutes, which was about three times as long as usual in large earthquakes. Because of the protracted ground shaking, the effects were severe. More than 400 buildings were destroyed and approximately 10,000 people died.

The wing of a modern multistory hospital collapsed, killing patients and medical personnel. Several other multistory structures were also destroyed, including the Hotel Regio. One guest in the hotel plunged five floors and survived with broken ribs. Another was on the third floor of an eight-story building when it collapsed and also survived because it collapsed backwards away from him.

A guest at the modern Sheraton Hotel was staying on the tenth floor and remembered the building swaying as much as 5 or 6 feet back and forth. Yet it did not collapse. The damage, as so often seems the case, was selective. While portions of the city felt only mild ground shaking and experienced little damage, the result was different on the section built over the old lake beds. Undamaged buildings stood next to collapsed structures (Fig. 8.22).

When ground shaking ceased and the dust settled, the body count began. Approximately 10,000 people had died, with 30,000 injured and 100,000 homeless. More than 400 structures were destroyed and another 3200 suffered damage, all of this in a city 220 miles from the epicenter of the quake. Urban areas close to the epicenter had, in fact, sustained relatively less damage.

One reason Mexico City suffered so much from a distant tremor was because much of the city was built over the sediments and muds filling in an old lake. It was in this

Figure 8.22   The selective destruction of buildings in the 1985 Mexico City earthquake.

part of Mexico City that the damage was most severe. The experience had been the same in the San Francisco Bay area as a result of the 1989 Loma Prieta earthquake. The Bay Area was also distant from the epicenter, but the response of loose sediments magnifies the effects of ground shaking. In both tremors, this led to considerable damage at large distances from the source.

Examination of the structures damaged in the Mexico City earthquake revealed a second important factor related to hazard. Many of the buildings that failed were multistory structures between 6 and 15 stories high. These buildings had responded like inverted clock pendulums, and they collapsed because of violent back-and-forth motions. Although most were well-built structures, buildings in this range of height had the unfortunate property of oscillation, or swing, of about 2 to 3 seconds—unfortunate because the earthquake, really two shocks close together in time, had generated an unusually long train of waves with a period of 2 seconds between wave crests. When these waves entered the lake beds under Mexico City they were actually increased in amplitude and shook down the high-rise buildings. This reaction of the buildings with their strong pendulumlike swings is termed *resonance catastrophe.*

Lesson: Mexico City is yet another example of an urban area located in a very unsuitable place. Unfortunately, level or flat terrain is easy to build on, but is also frequently underlain by soft sediments, which amplify ground shaking and lead to severe damage and building failure. The only solution is to learn from tremors such as the 1985 event, and to improve building design to compensate.

SUMMARY

An examination of large damaging earthquakes through history can serve as an important learning experience. If we understand why lives are lost and buildings fail, this knowledge may be used in the future to eliminate the causes, or at least to reduce the magnitude of loss.

The examples found in this chapter show that earthquake loss is due to several causes. The first is building collapse. As suggested: "Earthquakes don't kill people; buildings kill people." This is a bit too shortsighted, but certainly the failure of structures is a major cause of loss of life. At Kourion on Cyprus the collapse of unreinforced masonry (block) buildings triggered most of the deaths. So, too, at Lisbon in 1755, the collapse of churches built of unreinforced masonry contributed greatly to the death toll. Even well-designed reinforced structures, such as the high-rise buildings at Mexico City, can collapse under extreme conditions.

But it is not just building failure that causes loss of life in earthquakes. A number of related effects can occur when ground shaking begins. Fault offset can destroy structures built on top of the fractures. This was the case along the length of the 400-kilometer offset on the San Andreas fault in 1906. Today, the effects would be much worse given the high population density in California. Hundreds to thousands of structures are built atop the San Andreas fault, including public schools, hospitals, police and fire stations, and so forth.

Intense ground shaking of water-saturated sediments produces liquefaction, which was a problem at Port Royal, Jamaica, and in the 1995 Kobe, Japan, tremor, where much of the waterfront area liquefied, toppling buildings and disabling the port. A slide associated with liquefaction of an underlying layer occurred at Turnagin Heights during the

Alaska Good Friday earthquake. At present, liquefaction damage is difficult to avoid completely through building design precautions.

Common effects of earthquake ground shaking have been landslides and fire. Landslides of soft deposits at Shansi, China, in 1556 contributed to the disastrous loss of life and also caused casualties at Valdivia, Chile, in 1575 after a great earthquake there. A mountain stream had been blocked by an enormous landslide deposit that built up a temporary lake. When the water burst through the landslide, the torrent swept away everything that lay downstream from the slide. Nor should Yungay and Ranrahirca, Peru, be forgotten as they were buried in 1970 by a landslide triggered by an earthquake.

Fire is the easiest secondary effect of earthquakes to design for and to prevent. It can also be the deadliest, as history shows. Fire was a major factor adding to the magnitude of the disasters at Lisbon, San Francisco, and Tokyo. Certainly improved building-construction techniques, with fewer flammable materials, and more sprinkler systems are design elements that can reduce the risk of fire. Equally important is the protection of municipal water storage and delivery systems, as was learned from the 1906 San Francisco quake and fire.

KEY WORDS

Alaska	liquefaction	resonance
Basel	Lisbon	San Francisco
Chile	loess	Shansi
fire	Mexico City	Sonora
Jamaica	New Madrid	Tokyo
Kansu	Peru	tsunami
Kourion	Port Royal	
landslide	Reelfoot Lake	

C H A P T E R   9

Earthquakes in the United States

INTRODUCTION

The study of the effects of historic earthquakes can result in valuable lessons. However, another way of using history to increase personal safety is to understand not only where earthquakes may occur but also their frequency. California is well known as earthquake country, but damaging earthquakes have also struck Boston, Massachusetts, and Charleston, South Carolina.

This chapter surveys the seismicity of the continental United States, Alaska, and Hawaii. To accomplish this, the chapter is divided geographically into two principal sections: the western United States and eastern United States. The eastern United States includes all of the country from the Rocky Mountain front eastward to the Atlantic Ocean. The reason for this division is that these two parts of the nation show a distinct difference in the character and frequency of earthquakes, the earthquake setting, and the response of the ground to the release of earthquake energy. The information provided in this chapter should be a valuable tool in evaluating personal risk in a geographic sense.

THE WESTERN UNITED STATES

THE PLATE BOUNDARY: CALIFORNIA, OREGON, WASHINGTON, AND ALASKA

Earthquakes occurring along the Pacific coast states can be directly related to the presence of plate tectonic boundaries (Fig. 9.1). All three kinds of plate boundaries occur along the western boundary of North America, and California sees the effects of all three. South of California, in the Gulf of California, an oceanic ridge-fracture system marks the boundary where expansion of the crust is occurring between the North American and Pacific tectonic plates. The concentration of force along the plate boundary results in numerous earthquakes. The boundary provides a threat to California because it continues under southern California, providing local heat flow beneath the surface, and surficially expressed by large fault systems, such as the Imperial fault (Fig. 9.2). Evidence of the hazard posed can be seen by noting the effects of the 1940 and 1979 Imperial Valley earthquakes. The 1940 tremor was magnitude 7.1 and caused extensive damage as far east as Yuma, Arizona. The 1979 event resulted in the loss of the new Imperial County building and left evidence of faulting across the countryside (Fig. 9.3).

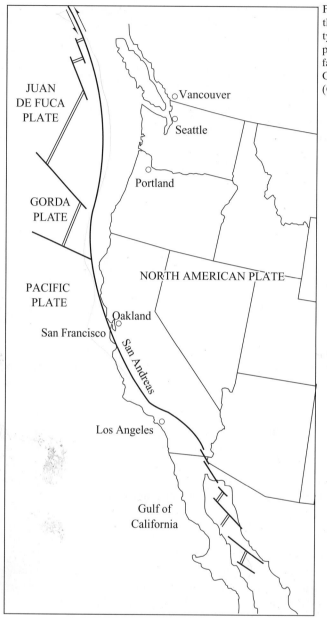

Figure 9.1   Plate tectonic setting of the western United States. All three types of plate boundaries are present: transform (San Andreas fault), ocean ridge (Gulf of California), and convergent-trench (Gorda plate boundary).

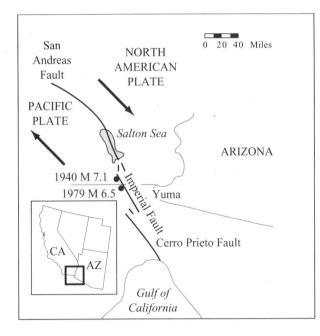

Figure 9.2   Plate boundary faults in southern California and northern Mexico: Imperial, Cerro Prieto, San Jacinto.

North of the Imperial Valley, evidence of buried oceanic ridges disappears and the plate boundary is centered on a large fault zone, the San Andreas, which stretches north past San Francisco to Cape Mendocino. This fault system is continuously active, producing the highest rate of seismicity on land in the lower 48 states. This activity includes some of the largest earthquakes in U.S. history, such as the 1857 Fort Tejon tremor and the 1906 San Francisco earthquake, both of which were greater than magnitude 8. The San Andreas fault zone runs close to or beneath many communities (Fig. 9.4). Clearly, the largest community at risk along the fault is San Francisco, because the fault passes directly through large portions of the San Francisco urban area (Fig. 9.5).

Oakland, to the east across San Francisco Bay, does not have to worry as much about the San Andreas fault, but it has a problem of its own. Passing through the heart of Oakland's heavily urbanized area is a principal branch of the San Andreas, the Hay-

Figure 9.3   Rows of crops in the Imperial Valley, California, offset by a fault slip that occurred during the 1979 earthquake.

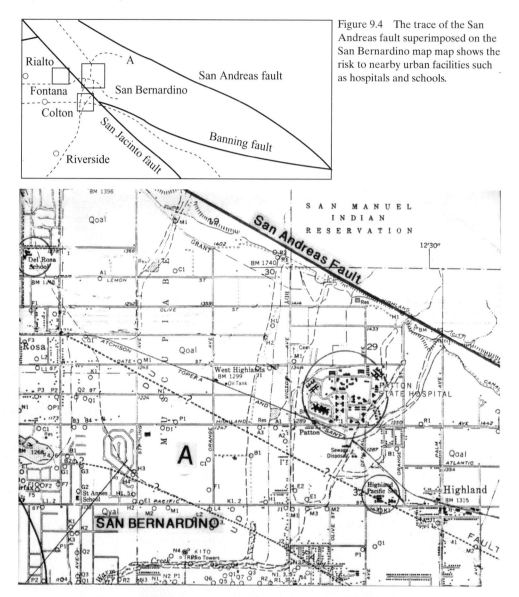

Figure 9.4 The trace of the San Andreas fault superimposed on the San Bernardino map map shows the risk to nearby urban facilities such as hospitals and schools.

ward fault. Numerous structures and facilities sit directly astride the fault zone, including the football stadium for the University of California (Berkeley). Think of that the next time you see a Pacific 10 football game played there on television. The Hayward fault has generated potentially devastating tremors in the past.

The San Andreas fault swings to the west and offshore at Cape Mendocino, California. At that point it becomes known as the Mendocino fracture zone, connecting to an oceanic ridge, the Gorda ridge (Fig. 9.6). The hazard to northern California coastal communities comes not only from tremors on the Mendocino and San Andreas faults but from earthquakes generated by the breaking loose of the Gorda plate as it is forced

Figure 9.5   The trace of the San Andreas fault passes right through the densely urbanized San Francisco area.

or subducted beneath the crust of northern California. A secondary hazard from plate subduction is that of tsunamis generated by tremors offshore along the plate boundary. A small tsunami was created by the earthquake of September 1, 1994, a magnitude-7.1 event.

The presence of a large plate of rock being forced under the western edge of the United States continues to exist off the coasts of Oregon and Washington. This creates considerable hazard for cities like Portland, Seattle, Tacoma, and Vancouver, Canada. This is clearly seen in a cross section of Earth's crust, which includes Seattle, where the Juan de Fuca plate is being subducted (Fig. 9.7).

The Juan de Fuca plate stretches along the western coast of North America for 1000 miles (1600 kilometers) from northern California to Canada. At its area of contact with North America, there is the potential for very large plate-boundary earthquakes of magnitude Mw9.0. Although such events have not occurred since European settlement of the area, evidence from older raised shorelines suggests several have taken place in the last 2000 years. Such an event would be roughly equivalent to the 1964 Good Friday earthquake in Alaska.

The Juan de Fuca plate also breaks and slips inside itself, releasing earthquake energy. As the plate penetrates Earth's interior, stresses build up to the point where failure occurs in a zone at depths of 45 to 60 kilometers, directly beneath the Seattle-Tacoma area. The 1949 (Mb7.1) and 1965 (Mb6.5) tremors are examples of intraplate events.

The remaining seismicity in Washington State and British Columbia appears to result from stress in the crust of the North American plate generated by pressure against it from the subducting plates. This includes two events of magnitude 7+ in 1918 and 1946 as well as one large tremor in the northern Cascades of Washington in 1872.

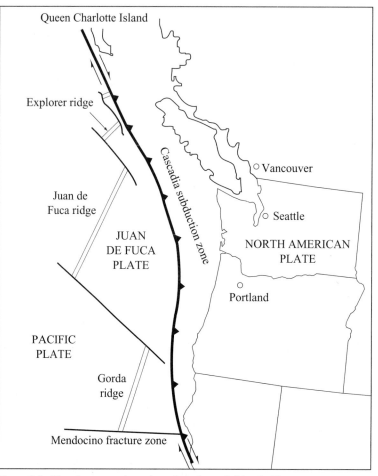

Figure 9.6 The plate tectonic setting of northern California. The movement of the Gorda plate under California generates potentially damaging earthquakes.

The Pacific tectonic plate forces its way under the North American continent along the southern coastline of Alaska. This occurs along a 2800-mile arc stretching from southeast of Anchorage out to the west along the Aleutian Island chain reaching almost to Asia (Fig. 9.8). The earthquakes that result are frequent and often severe. It was this plate motion that resulted in the enormous energy release of the 1964 earthquake in Alaska and its aftershock sequence. As the Pacific plate stretches to the north under Alaska it creates a zone of earthquake activity whose foci extend northward underneath the state through one of the more habitable regions.

The peninsula of Alaska, including the capital Juneau, faces moderate earthquake hazard from the Fairweather and Queen Charlotte Islands fracture zone, which forms the plate boundary in the Pacific Ocean to the west (Fig. 9.8). Sitka, Alaska, was rocked by a magnitude-7.3 tremor in 1972, which did minor damage. Only the central and northern regions of Alaska, much of which are only marginally habitable in the winter, are relatively free of earthquakes.

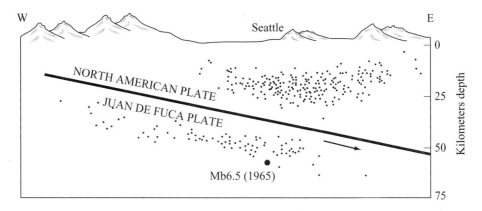

Figure 9.7   Eastwest cross section through Earth at the latitude of Seattle, Washington. Locations of earthquake foci are indicated by dots.

THE BASIN AND RANGE: ARIZONA, NEVADA, UTAH, CALIFORNIA, MONTANA, IDAHO, AND OREGON

Earth's crust across much of the interior of the western United States is slowly extending, increasing east-west in width. Earthquake energy is being released as a result of this action. The faulting associated with the earthquakes has, over time, resulted in a distinctive alternating valley and mountain topography. Consequently, this region is termed the *Basin and Range*. Occasional great earthquakes will occur, as at Owens Valley in 1872, when an estimated magnitude-8+ tremor shook eastern California (Fig. 9.9).

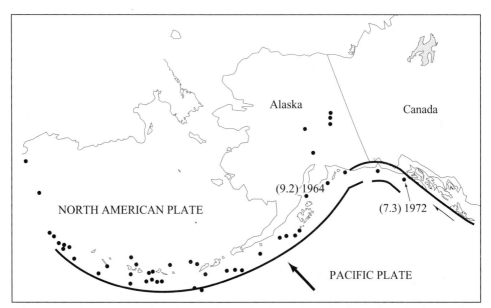

Figure 9.8   Plate tectonic setting of Alaska. This region is dominated by the northward underthrusting of the Pacific plate beneath the North American plate.

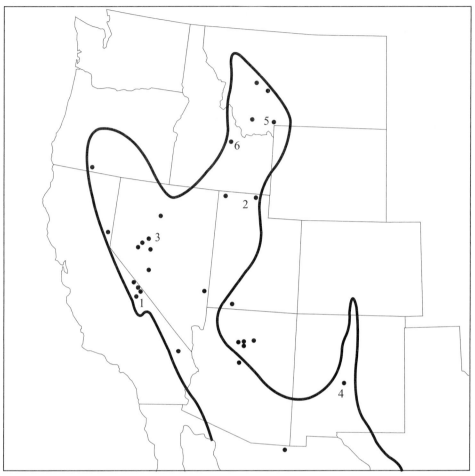

Figure 9.9 Great Basin and Basin and Range of the western United States. Large historic earthquakes are shown. Key: 1 = 1872, Owens Valley; 2 = 1934, Hansel Valley; 3 = 1954, Fallon tremors; 4 = 1906, Socorro; 5 = 1959, Hebgen Lake; 6 = 1983, Borah Park.

Eastern California remains sparsely populated today. However, central Utah, which lies along the eastern Basin and Range boundary, is more densely populated. Along a narrow corridor stretching north-south lie Ogden, Salt Lake City, and Provo, Utah's principal cities. These urban areas lie astride the Wasatch fault zone, which has generated damaging quakes several times in the last 150 years. The Hansel Valley earthquake of 1934 on the Utah-Idaho border caused minor damage to the south in Salt Lake City. This was a gentle reminder of potential danger in the future.

The earthquake activity associated with the Wasatch fault continues to the north into Idaho and Montana. This belt of seismicity has been termed the *Intermountain Seismic Belt* (ISB) and has the highest rate of seismicity of any region in the United States east of Nevada.

Earthquake activity in Montana extends over a broad area in the western part of the state. Much of the activity is due to the extension of the northern end of the ISB into this region (Fig. 9.9). However, two zones of faulting and seismicity extend across Montana: the Centennial Tectonic Belt and the Lewis and Clark Zone. The destructive 1935 earthquakes near Helena, Montana (Ms6.2, 6.0), occurred at the eastern end of the Lewis and Clark Zone. Two large earthquakes have occurred since 1900 in the Centennial Tectonic Belt: the 1983 (Ms7.3) Borah Peak tremor and the 1959 Hebgen Lake event (Ms7.5).

The Hebgen Lake event was the largest Montana tremor since 1900 and was felt over an area of 500,000 square miles, reactivating 160 geysers in Yellowstone National Park. Clearly the worst effect from this tremor was extensive landsliding in the region. The worst slide in the Madison River valley buried 28 people and dammed up the Madison River for a time, creating Quake Lake. This locale has been designated a national landmark, and the scars from the landslide can still be seen on the mountains.

The Borah Peak and Hebgen Lake tremors represent normal faulting as a result of stretching of Earth's crust. The Borah Peak, Idaho, earthquake created a fault scarp along 36 kilometers (22.5 miles) of the trace of the Lost River fault and triggered building collapse in Chalfant, Idaho (see Box 3.1). Two people were killed in this earthquake from building collapse.

One other Montana tremor deserves mention because of its size. This is the 1925 Clarkston Valley earthquake of magnitude 6.75. Unlike the Borah Peak and Hebgen Lake tremors, the Clarkston Valley earthquake left no prominent fault scarps at the surface. The activity in Idaho and Montana suggests a potential for devastating magnitude-7.0 tremors elsewhere along the Intermountain Seismic Belt.

Far to the south of Idaho and Montana, Arizona has a concentration of earthquake activity in the northern part of the state. Northern Arizona shook from three moderate earthquakes early in the twentieth century. The 1906 (Mb6.2), 1910 (Mb6.0), and 1912 (Mb6.2) quakes affected what at the time was a very sparsely populated region. A reminder was added in 1993 when a Mw5.3 tremor woke up numerous residents of Flagstaff, Arizona, and knocked out power for a short time at the Grand Canyon (Fig. 5.17). Grand Canyon National Park faces significant hazard from earthquake-triggered landslides from the many steep slopes within the Grand Canyon. Rockfalls and slides have been triggered from tremors as small as $M_L 4.5$.

Nevada is considered by the federal government to have an extreme level of earthquake hazard. This is because of the numerous Basin and Range-type faults located in the state, many of which appear to be active. A map of epicenter activity for Nevada shows the frequency of activity in historic time (Fig. 9.10). The densest concentration of epicenters can be seen in the western part of the state, with a smaller concentration to the south, the so-called Nevada Seismic Zone. The seismic zone is the southern extension of the Intermountain Seismic Belt, whereas much of the activity in western Nevada is a result of the presence of the western boundary of the Basin and Range.

Nevada has experienced a number of large tremors since 1900. Perhaps the best documented was the sequence that occurred in July, August, and December 1954 in west- central Nevada in the vicinity of Fallon (Fig. 9.11). The first tremor occurred on July 6 (Mw5.9) followed by another Mw5.9 aftershock. About 10 miles farther north on August 23 an Mw6.5 tremor occurred. Nor was this the end of it, for on December 16, about 25 miles east of the July 6 tremor, an Mw7.2 shock hit Fairview Peak, followed 4 minutes later by an Mw6.7 event. The December 16 shocks alone resulted in 65 miles of surface faulting, with scarps up to 12 feet high (Fig. 9.12). Fortunately, these tremors occurred

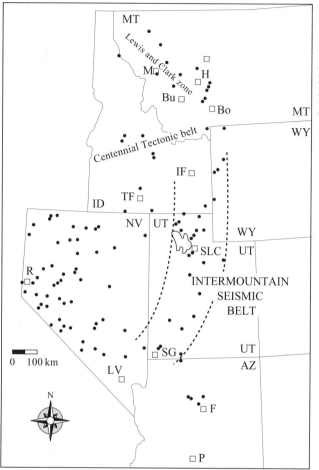

Figure 9.10 Representative seismicity (M≥5.0) in the mountainous interior of the western United States. Key: Bo = Bozeman; Bu = Butte; F = Flagstaff; H = Helena; IF = Idaho Falls; LV = Las Vegas; M = Missoula; P = Phoenix; R = Reno; SG = Saint George; SLC = Salt Lake City; TF = Twin Falls.

in a sparsely populated area, and only minor damage resulted. There would be reason for concern in Nevada if such events or sequence of events occurred under the urban Reno-Carson City corridor.

A large Basin and Range earthquake took place in 1915 in north-central Nevada. This was magnitude Mw6.9 and was felt from San Francisco to Salt Lake City, and from Oregon to San Diego. Once again, disaster was averted, for this tremor occurred in the sparsely populated Pleasant Valley, creating fault scarps up to 15 feet high (Fig. 9.12). Events like this one indicate the potential that Basin and Range faults have for producing large and potentially damaging earthquakes.

THE RIO GRANDE RIFT: TEXAS, NEW MEXICO, AND COLORADO

Stretching from central Colorado to western Texas, the Rio Grande has cut a large gorge into Earth's crust. This has been made easier by the fact that along the course of the river the crustal rocks have been broken and stretched. The broken rock has a tendency to fill up the increased space from stretching, forming a depression. Such a depression

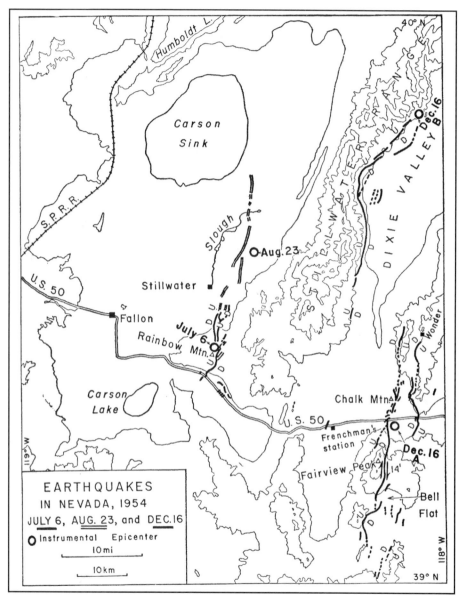

Figure 9.11   Map of Fallon, Nevada, area showing epicenters of 1954 quakes and the surface fault traces that broke the ground.(From *Elementary Seismology* by Richter, ©1958 by W. H. Freeman and Company. Used with permission.)

is termed a *rift*. As the rock is stretched and breaks, earthquake energy is released. Thus, earthquake hazard exists for cities like Santa Fe, Albuquerque, and El Paso (Fig. 9.13).

The year 1906 was a busy one for the western United States. Not only did San Francisco rock to its great earthquake, but Arizona had its 1906 event (Ms6.2) and New Mexico shook to a sequence of its largest historic tremors. Socorro, New Mexico, lies in the Rio Grande Rift about halfway between Albuquerque and El Paso. On July 2, 1906,

Figure 9.12   Fault scarp created during the Dec. 16, 1954, Nevada earthquake.

Socorro felt a tremor that reached Modified Mercalli intensity VIII (approximately Richter magnitude 6.1). This was followed by two even more severe shocks on July 16 (M6.3) and November 15 (M6.5).

The northern and southern ends of the Rio Grande Rift in Colorado and Texas, respectively, do not appear to be overly active. The August 16, 1931, Valentine earthquake is the only notable historic event in west Texas associated with the spreading of the Rio Grande Rift/Basin and Range. This event was due to normal faulting and had a calculated magnitude of Mw6.3.

The northern extension of the Rio Grande Rift into Colorado also appears to be without large significant events in historic time. The closest event outside the rift would be the 1966 Dulce, New Mexico, tremor on the Colorado border in 1966. The largest historic event in Colorado would instead have to be the November 7, 1882, tremor that, based on ground-shaking reports, had an epicenter just west of Fort Collins. This tremor was apparently a result of release of stress along a mountain-front fault at depth and was estimated to have a magnitude of approximately Mw6.6. Such events are rare throughout the Rocky Mountains, and at this point unpredictable in occurrence.

HAWAII

The Hawaiian Islands lie well inside the Pacific tectonic plate, far away from the plate boundaries where stresses are more frequently released in the form of earthquakes. Hawaii is well known for its volcanic eruptions, yet it also has occasional damaging earthquakes. Most of these earthquakes are related to stresses associated with volcanic activity.

The Hawaiian Islands lie above a *hot spot* located within the Pacific plate. Hot spots exist at what appear to be random locations and their origin is not well understood. What is known is that heat is rising from Earth's interior at these locations, and at shallow depths (10 to 15 kilometers beneath the surface) is associated with hot liquid

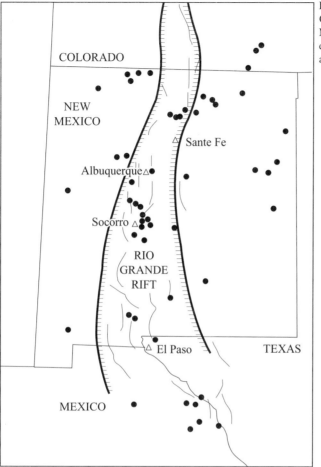

Figure 9.13   A map of the Rio Grande Rift of Colorado, New Mexico, and Texas. Selected epicenters of earthquakes associated with the rift are shown.

rock material called *magma*, which is the source of the lavas that are produced in volcanic eruptions.

As magma tries to rise buoyantly to the surface it forces its way through the overlying rock, causing the rock to break, resulting in earthquakes. Such earthquakes occur frequently in the rock surrounding the magma, sometimes shaking the ground so often that it is nearly continuous, giving rise to the term *harmonic tremors* (Fig. 9.14).

Large tremors occur less frequently in response to swelling of the volcanic cone as it fills with magma. The 1975 magnitude-7.5 Kalapana earthquake occurred as stress was released on a thrust fault at the base of Kilauea volcano. This thrust was located at a depth corresponding to the base of the volcano, which suggested movement of Kilauea's south flank seaward (Fig. 9.15).

Probably 95 percent of Hawaiian earthquakes are related to the volcanic activity of the hot spot. However, some tremors are not volcanic in origin but related to the Molokai Fracture Zone, which passes offshore of Oahu and Honolulu, so close in fact that it is only miles away (Fig. 9.16). It has been estimated that the Molokai Fracture

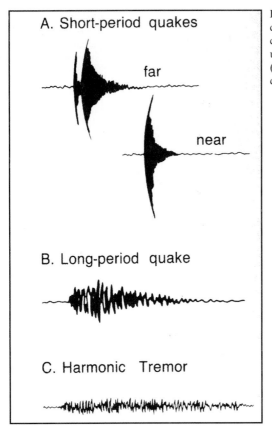

A. Short-period quakes

far

near

B. Long-period quake

C. Harmonic Tremor

Figure 9.14 Seismogram of nearly continuous harmonic tremors caused by magma movement underground prior to an eruption (c), compared to nonvolcanic quakes.

Zone is capable of producing a magnitude-7.0 earthquake, and is thus a major hazard for Honolulu.

THE EASTERN UNITED STATES

NEW ENGLAND

Nantucket Island was jolted by a small tremor on October 24, 1965. Many residents were surprised to feel an earthquake in Massachusetts. Yet this was not the first, as a shock of Modified Mercalli intensity VIII shook eastern Massachusetts in November 1755. At Boston, walls and chimneys were thrown down, and here and there stone fences collapsed. One observer reported that "the earth seemed to wave like the waves of the sea." Nor was this the only large shock to affect Massachusetts; there were others in 1638 and 1663 that damaged chimneys at Plymouth, Lynn, and Salem. Elsewhere, New-bury was strongly shaken in 1643, and eastern Massachusetts shook several times in the eighteenth century: 1727, 1737, 1741, and 1744.

Massachusetts, as is much of the rest of New England, is subject to earthquake shocks from two different source areas. To the north, under the Saint Lawrence River valley of Canada lies an ancient zone of weakness and faulting. The 1663 Three Rivers tremor was located along the lower Saint Lawrence and reached a maximum Modified

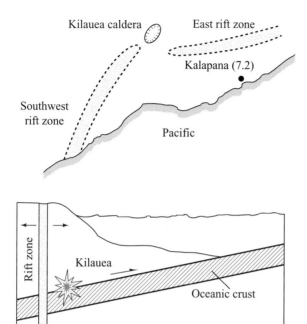

Figure 9.15   Kalapana earthquake caused by swelling of the south flank of Kilauea volcano. The south flank failed by thrust faulting, releasing earthquake energy.

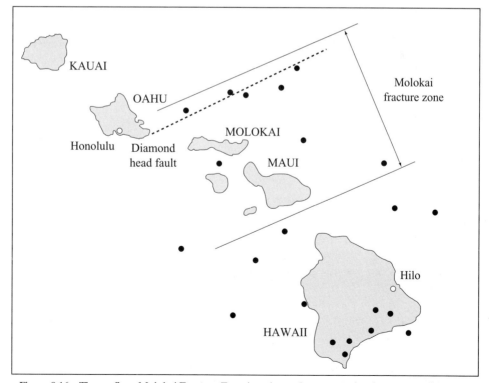

Figure 9.16   The seafloor Molokai Fracture Zone is active and poses a major threat to nearby Honolulu. Circles are earthquake epicenters.

Mercalli intensity of X. On October 20, 1870, the Saint Lawrence River valley shook from a Modified Mercalli IX shock. It has been suggested that an ancient zone of rifting underlies the Saint Lawrence River valley, which began to tear the crust apart, but did not succeed. Thus, the weakened crust today responds to stress by reactivating ancient buried faults.

The most severe historic earthquake experienced in Massachusetts was the 1755 tremor off Cape Ann, beneath the ocean floor. A plot of New England earthquake epicenters suggests the presence of a second source area (Fig. 9.17). Stretching from Boston to Canada, the zone of earthquake activity has been termed the Boston-Ottawa zone. These earthquake epicenters are also aligned with a linear trend of ancient submarine volcanic mountains, the Kelvin seamount chain. It has been suggested that these volcanic features are located along the trace of an ancient oceanic fracture, perhaps a transform fault, which may extend into the continental crust.

SOUTH CAROLINA AND THE APPALACHIANS

The most destructive earthquake to strike the southeastern United States was the August 31, 1886, tremor that rattled South Carolina. It was estimated to be Ms7.7, and was felt over an area of 5,000,000 square kilometers. The tremor was felt as far away as Detroit and Boston. A total of 60 people died, and property damage was extensive, especially in Charleston, South Carolina, which appears to have been close to the epicenter. More than 1500 buildings in Charleston were listed as damaged, out of a total of 7000 structures in existence at the time. Nearly all chimneys (14,000) were damaged or destroyed. No faulting was expressed at Earth's surface, and because this event occurred at the beginning of the era of the seismograph, we have no clear picture of the type of faulting or the source of the earthquake.

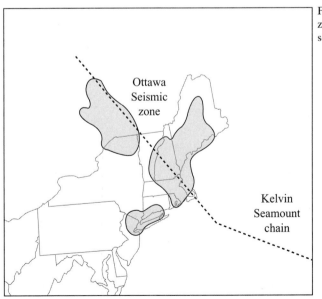

Figure 9.17 New England seismic zones (shaded) and the Kelvin seamount chain.

A diffuse zone of seismicity stretches across South Carolina, but it is weakly defined, and events are so infrequent that it is difficult to characterize and to relate to a specific fault zone or cause. Suggestions have been made that because of its more or less linear pattern it may be associated with activity along an extension of the offshore oceanic Blake Fracture Zone. This is by no means certain or widely accepted as a cause.

Some of the South Carolina seismicity extends into the southern Appalachians. Crossing this trend like the top of the letter T is the Southern Appalachian Seismic Zone. The epicenters are largely confined to the rolling valley and ridge country extending from northern Alabama to western Virginia and includes cities such as Chattanooga, Knoxville, and Bristol, Tennessee, as well as Roanoke, Virginia (Fig. 9.18). The cause of the seismicity is not well understood, consisting mostly of minor tremors. An occasional damaging tremor has occurred in this seismic zone.

On May 31, 1897, a Modified Mercalli VIII tremor shook northwestern Virginia and was felt from Pennsylvania to Georgia. Minor damage was reported as far away as Raleigh, North Carolina, and Bristol and Knoxville, Tennessee. The largest historic tremor in eastern Tennessee was a Modified Mercalli VII event on March 28, 1913. Centered near Knoxville, the damage was minor and the area where it was felt was confined to eastern Tennessee. Observers reported the ground rising and falling.

MISSISSIPPI VALLEY

The Mississippi River valley region may well have the greatest seismic risk of any area of the United States east of the Rocky Mountains. The Mississippi River valley and the lower Ohio River valley are underlain by an aborted rift zone, much like that envisioned for the lower Saint Lawrence. Small earthquakes occur along the zone frequently, often reaching magnitude 4 (Fig. 9.19). An occasional damaging event, of moderate size, will occur. On January 4, 1843, a Modified Mercalli VIII tremor occurred near Memphis, Tennessee, causing minor damage. This shock was felt over an area of 1,000,000 square

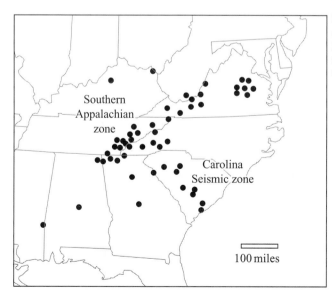

Figure 9.18   Seismicity in the southeastern United States is largely centered in South Carolina and along the southern Appalachian seismic zone.

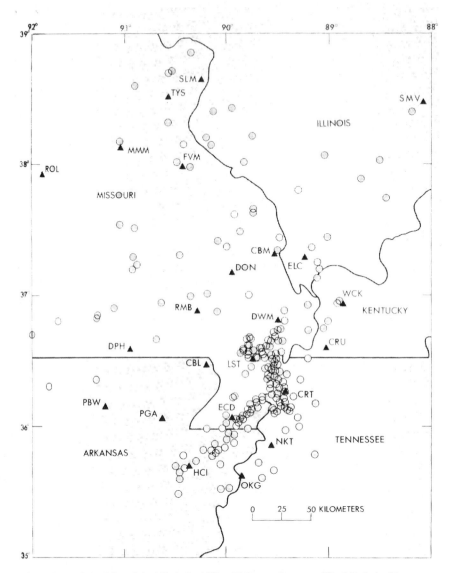

Figure 9.19 Seismicity of the Mississippi River Valley region open. The Mississippi is underlain by the active faults of an ancient rift system. Open circles are epicenters.

kilometers. Similar shocks ranging from intensity V to VII have been centered in western Tennessee along the zone of weakness in 1889, 1941, 1952, 1955, and 1956.

The real concern, however, would be the much less frequent major tremors, such as the 1811 to 1812 sequence that some believe to be among the largest to strike North America in historic time. The three main shocks have been estimated to have been at least magnitude 8 and possibly larger. The recurrence of such tremors today would have a devastating impact (Fig. 9.20). Heavy damage could be expected in Memphis and Saint

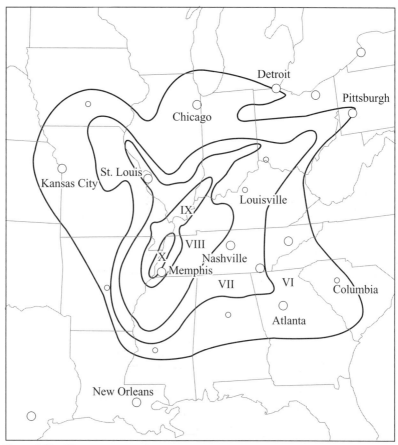

Figure 9.20   Worst-case scenario of ground shaking for recurrence of the 1811–1812 size tremors along the Mississippi rift.

Louis as well as in smaller communities in the region. Significant damage could be expected as far away as Little Rock, Cincinnati, Louisville, Nashville, and Chicago.

SUMMARY

The most active seismic areas in the United States are the result of activity along plate tectonic boundaries. This accounts for the bulk of the activity in the western United States in California, Oregon, Washington, and Alaska. The San Andreas fault marks the boundary of the North American and Pacific plates in California, and movement of the plates along this boundary has accounted for the largest tremors in 1857 (Fort Tejon) and 1906 (San Francisco). Such tremors will continue to occur as a result of plate movement.

Large earthquakes will also occur to the north in Oregon and Washington where the plate-boundary activity consists of the Juan de Fuca plate moving beneath the North American plate. The 1949 and 1965 tremors in Washington State were a result of this action. The threat of future large earthquakes along this boundary is very real for Portland, Seattle, and Vancouver, Canada.

The earthquake threat to Alaska exists from both San Andreas-type plate-boundary faults (e.g., Fairweather), which can generate magnitude-7.0 tremors, and the underthrusting of the Pacific plate beneath North America along much of the Alaskan coastline. The latter type of activity was responsible for the great Alaskan earthquake of 1964 (Mw9.2), which rocked Anchorage and devastated communities along the Alaskan coastline.

Separate from the plate-boundary events in the western United States are the tremors in Hawaii and in the intermountain states. Hawaii has large tremors either directly or indirectly related to volcanic eruptions. The swelling and expansion of Kilauea volcano has produced tremors as large as magnitude 7.5. In addition, activity in fracture zones on the ocean floor poses a threat to Honolulu.

Earthquakes as large as magnitude Mw7.3 have struck the intermountain western states. Such events occur in response to expansion of Earth's crust across an area stretching from eastern California to central Utah, and from Canada to Mexico. This poses a threat for communities such as Reno, Salt Lake City, and Boise.

The eastern United States has a much lower level of seismic activity, which makes it more difficult to assess the seismic threat. Zones of seismic activity have been identified, but causes are much less certain. This seismic activity is termed *intraplate*—that is, seismicity occurring well within the North American plate and far away from plate boundaries. Much of this activity appears to be related to activity on buried faults that are part of ancient aborted rift zones, where the crust began to separate and then stopped. This would appear to be the cause of activity along the Mississippi and Ohio River valleys, and the lower Saint Lawrence as well. Occasional large earthquakes, such as the 1811 and 1812 New Madrid tremors along the Mississippi and the 1870 Saint Lawrence shock, suggest the magnitude of the threat to cities like Memphis, Saint Louis, Quebec, Montreal, and Boston.

Significant activity along the eastern seaboard near Boston, Massachusetts, and Charleston, South Carolina, is not well understood. They may be related to large ancient oceanic fractures. Whatever the cause, these locales have a history of such events, and thus they can be considered areas with significant earthquake hazard.

KEY WORDS

Albuquerque	Grand Canyon	New Madrid
Basin and Range	Honolulu	plate boundaries
Blake fracture zone	Imperial fault	Pleasant Valley
Boston-Ottowa zone	Intermountain Seismic Belt	Queen Charlotte Islands
Calaveras fault	Juan de Fuca	Rio Grande Rift
Cape Mendocino	Kelvin seamounts	Saint Lawrence
Charleston	Kilauea	San Andreas
Fairview Peak	Mississippi valley	Socorro
Fallon	Molokai Fracture Zone	southern Appalachians
Gorda ridge	Nevada Seismic Zone	Wasatch fault

Earthquake Prediction

INTRODUCTION

I can predict an earthquake—no, really I can. Somewhere in the world tomorrow there will be an earthquake. Now that wasn't too hard, was it? The only problem is that it's not much use. In fact, prediction isn't really the right word to use here. For a prediction to be useful it should stipulate the time, place, and size of an earthquake. If anyone could do that, he or she would be rich, but of course it cannot be consistently accomplished, at least not yet. Actually, a few earthquakes have been forecast. Note that use of the word *forecast* is a lot less specific, analogous to weather forecasting, which is not always precise.

SNAKES, YAKS, AND COCKROACHES

Any prediction or forecast must rely on things that happen before the event. The term to be used is *precursor*. Ever since ancient times the Chinese have relied upon animal behavior as an indication of a coming tremor. Any unusual behavior, usually from several groups of animals, was felt to be an indication that a quake was imminent.

An example of where animal precursors seemed to be useful was in warning of the February 4, 1975, Haicheng, China, earthquake. The first reports of anomalous animal behavior occurred in December 1974. The most striking report is that of snakes crawling from their holes in the ground and freezing on the surface. This is all the more unusual because snakes are normally in hibernation that time of year. Shortly after the snake report, an earthquake occurred with a maximum tremor of magnitude 4.8 in the sequence. It happened about 70 kilometers (44 miles) northeast of Haicheng.

During January 1975 numerous reports of anomalous animal behavior were received. This included snakes, rats, pigs, birds, fish, horses, cows, sheep, yaks, cats, and dogs. Dogs would bark ceaselessly, and restless behavior was exhibited by the other animals. This included birds circling and not roosting, animals not entering barns, and horses running wildly about. Yaks would not eat and rolled over repeatedly.

Unusual animal behavior was reported to have intensified just a few days before the February 4 tremor. Although the animal behavior was deemed by Chinese officials as an important clue to an impending tremor, the successful prediction, which was issued at least 5 hours before the earthquake, was based on geological and geophysical precursors as well.

Outside of China anomalous animal behavior is generally not taken seriously as an earthquake precursor. The reason is that it has not been possible to pin down conclusively the cause of such behavior. Many ideas have been advanced, but little research has been carried out on animals. The principal problem, of course, is given that animals sense something, it is not at all certain what is being sensed. Furthermore, a wide variety of different animals appear to be affected, with often little in common physiologically. Nor are the animals always successful predictors. Only scattered anomalous behavior of animals was reported before the disastrous 1976 Tangshan tremor, which resulted in 250,000 deaths.

NEW MADRID: A FALSE ALARM

Thus, although a few earthquakes have been successfully predicted it is the hit-and-miss record so far that has been a problem. Not only have other tremors not been predicted, but some serious false alarms have been raised. The best publicized was the December 1990 New Madrid (Missouri) prediction. This prediction, or *projection* as the predictor termed it, was made by a private citizen, Iben Browning. Browning was an inventor/lecturer with a doctoral degree in biology who had some influence in the business community, to which he frequently delivered lectures on climate, volcanoes, and earthquakes.

Browning believed he had discovered a relationship between the tidal pull of the Sun and Moon on the rocks in Earth's crust and the occurrence of volcanic eruptions and earthquake activity. Browning's idea was that periods of high tidal stress on the rocks of the crust could cause earthquakes in regions of existing stress much like a trigger mechanism. Browning identified regions of high earthquake risk by determining which bands of latitude on Earth would be subject to large amounts of rock tidal stresses, and then he would turn to the seismological literature to identify specific faults indicated by earthquake researchers as being potentially hazardous.

This idea was the basis of Browning's prediction of a New Madrid earthquake, which was to occur on December 3, 1990. This was a time period of high tidal stress because of the alignment of the Sun and Moon and their combined gravitational pull on Earth. Browning was also aware of the recent research on the New Madrid earthquakes of 1811 and 1812 and the Reelfoot rift zone in the Mississippi River valley. It must be pointed out, in all fairness, that Browning made no initial efforts to widely publicize this prediction. He discussed it in lectures given to business groups. It was left to others to spread the word, which ultimately reached the ears of the media, at which point it became grist for the mill, a media earthquake.

There were several problems with Browning's prediction. First, the idea was not a new one to the seismological sciences, but indeed had been tested many times and found wanting. The earliest attempt to relate tidal stresses to earthquakes was apparently in the mid- to late-1800s by Alexis Perrey. Subsequent studies, as late as the 1960s to 1980s, have almost without exception reported negative results.

Also a problem in almost all studies of the tidal stress–earthquake concept is one of a lack of correlation in time. It is necessary only to mention the fact that the New Madrid region has been subject to periods of high tidal stress in December 1964, December 1982, and between 1812 and 1814, without significant tremors occurring.

The Browning prediction was strongly opposed by the seismologic community. Aside from individual interviews with scientists who gave no credence whatsoever to the prediction it was also strongly condemned in a report released by the ad hoc working group on the December 2–3, 1990, prediction. The Working Group was a panel of scientists selected by the United States Geological Survey to study the prediction. The results of their study also discounted the prediction for many of the reasons already outlined. Nevertheless, because of media attention given the prediction, many in responsible positions, as well as many of the populace in the area, were strongly affected. The social and psychological effects of Browning's prediction were immense. The lives of hundreds of thousands of people in the central United States were disrupted, and tens of millions of dollars were needlessly expended. Earthquake preparedness officials, police, emergency response teams, and the National Guard were especially busy.

State earthquake preparedness officials in Missouri, Arkansas, Tennessee, and Kentucky were deluged with requests for information. State earthquake response budgets were drained to the tune of $200,000 by the need to prepare for the predicted event. This expenditure included establishment of emergency shelters, design of evacuation procedures, and conduct of earthquake drills. More than a thousand emergency response officials in Saint Louis participated in an earthquake mock scenario. This drill included firefighters, police, emergency medical technicians, and other emergency response people. National Guard troops in Kentucky, Missouri, and Arkansas were placed on a state of alert and participated in earthquake drills. These preparations and drills had a silver lining in that the preparedness of emergency response personnel was at the highest level it had ever been. However, much of the psychological impact on individuals was negative.

Many of the schools in the area were closed because of probable lack of attendance on December 3. New Madrid's largest business, Noranda Aluminum, with approximately 1500 employees, also shut down for the same reason. Other businesses suffered from employee absenteeism, some to the point of having to close. Many people left the area for a few days or as much as a week, often staying with relatives over the critical period, during which no earthquake occurred. It would seem unlikely that any of these people will be eager to take another earthquake prediction or warning seriously. The old adage of cry wolf applies here.

EARTHQUAKE PREDICTION: THE LONG AND SHORT OF IT

Tidal forces have so far not been useful in predicting or even forecasting earthquakes. In fact, no single phenomenon associated with impending tremors has been a reliable prediction tool. As part of the evolution of the concept, workers have also been looking at less precise and more long-term predictions, sometimes called *forecasts*, with some notable successes. Long-term prediction is usually concerned with events that may occur more than a year in the future. The term forecast is used because the time and place are less precise than in the way prediction has been defined. Such long-term predictions are usually based on a change in seismic activity or a lack of seismic activity in a region. In the latter case, the term *seismic gap* has been applied.

The Loma Prieta, California, earthquake of October 17, 1989, occurred in an area close to the San Andreas fault, which had been designated as a seismic gap. This was the famous World Series earthquake, which was actually viewed as it occurred on television across the U.S. For the Loma Prieta segment of the fault zone the U. S. Geological Survey had published a forecast of a 30 percent probability of failure in a magnitude

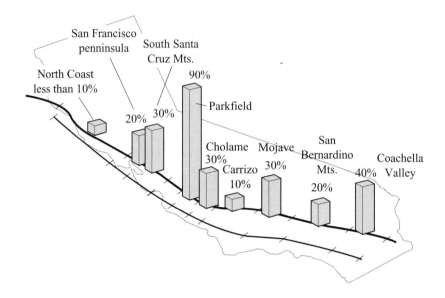

Figure 10.1   Probability forecast for earthquakes along the San Andreas fault zone.
Higher probability is indicated by taller rectangles.

≥ 6.5 earthquake within a 30-year time window. This is a fairly high probability when compared to the other segments of the San Andreas fault zone (Fig. 10.1). When the Ms7.1 Loma Prieta earthquake struck there were claims that it had been forecast as the tremor and its aftershocks seemed to fill a gap in seismicity along the San Andreas fault (Fig. 10.2). However, some have disagreed because of a lack of surface displacement along the trace of the San Andreas, maintaining that it must have been another nearby fault.

There are other cases of long-term predictions based on seismic gaps, some of which are in the wait-and-see category. The seismic-gap approach has been successfully used in forecasting at least ten large plate-boundary earthquakes. For example, before the 1978 Oaxaca, Mexico, earthquake (magnitude 7.8), there had been a quiet interval of 5 years. Renewed activity began 10 months before the tremor and was the basis of the forecast.

There are a number of seismic gaps around the Pacific Rim. The U.S. Geological Survey has kept a close watch on seismic gaps along the Aleutian Island chain. Because the Pacific/North American plate boundary is located there, a high potential exists for gaps being filled by large and damaging tremors similar to the 1964 Alaskan earthquake (Fig. 10.3).

The 1975 Haicheng earthquake would be a good example of a tremor where a long-term forecast as well as shorter-term predictions were based in part on changes in seismicity. This tremor has already been noted because of the role of anomalous animal behavior in prediction. However, physical precursors, such as seismic activity, were also used. In fact, the earliest long-term predictions were based on seismic activity. A long-term prediction for Liaoning Province, in which Haicheng is located, was first issued

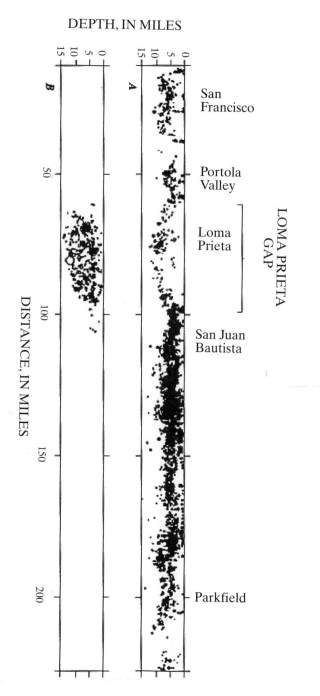

DEPTH, IN MILES

DISTANCE, IN MILES

San Francisco

Portola Valley

Loma Prieta

LOMA PRIETA GAP

San Juan Bautista

Parkfield

Figure 10.2   View of the surface of the San Andreas fault showing the gap in seismicity on the Loma Prieta segment of the fault before the 1989 tremor.

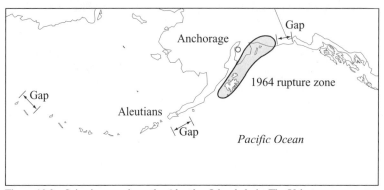

Figure 10.3 Seismic gaps along the Aleutian Island chain. The Yakataga gap east of Anchorage is of special concern because of its proximity to the city.

in 1970 and based almost solely on the northerly migration of larger tremors toward the area, and a general increase of regional seismicity. This level of seismicity continued to increase markedly in December 1974 with a swarm of tremors capped by a magnitude-4.8 event. A short-term prediction was issued on January 13, 1975, for a magnitude-5.5–6.0 event in the northern Liaotung Peninsula during the first half of 1975 (Fig. 10.4). This revised prediction was based on many precursors, including ground tilt, changes in groundwater content, Earth's resistivity, anomalous animal behavior, and, of course, anomalous seismicity patterns.

The final prediction was issued on February 4, 1975, the day of the earthquake, and seems to have been based primarily on the end of a swarm of small tremors that began on February 1 at the same locale as the main shock. This prediction was released 5 to 10 hours before the main shock, and was likely responsible for saving countless

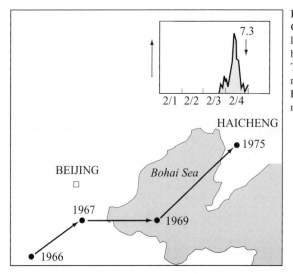

Figure 10.4 A map of northeastern China showing the migration of large seismic events (i.e., M5–6) before the 1975 Haicheng tremor. The graph shows the change in number of foreshocks between February 1–4, before the 7.3-magnitude main shock.

thousands of lives, as people evacuated the buildings in Haicheng and the surrounding area. Most of these same buildings subsequently collapsed in the tremor or were severely damaged.

Thus, the 1975 Haicheng prediction marked a success. But as if to remind the celebrants of the difficulties inherent in earthquake prediction, this was followed by the completely unexpected 1976 Tangshan, China, tremor, which was *not* predicted and killed an estimated 250,000 people. There was a lack of anomalous activity before this latter event, and this points out the problem with using precursors as prediction tools. None of the potential physical precursors that have been identified are uniformly reliable in the majority of cases. In fact, even where a number of physical precursors are monitored, the results are often unsuccessful in predicting a tremor.

SHORT-TERM PREDICTION: PRECURSORS, SUCCESSES, AND FAILURES

Before rock failure and the release of earthquake energy, certain physical changes occur in rocks and at Earth's surface as rocks respond to the stresses along faults. Rock properties that change under stress include rock strength, electrical properties of rock, and seismic velocity in rocks. Laboratory experiments have shown this to be so. These changes are recognized by the presence of physical anomalies—that is, physical precursors. One example would be unusual seismic activity (e.g., the seismic gap).

As rock is compressed it may break and slip in weak areas, releasing earthquake energy. This may result in a sequence of earthquakes before the main shock. Such a sequence is termed *foreshock activity* and might include large numbers of small quakes known as *microearthquakes*. Foreshock activity can be a useful indicator of the larger tremor to follow, as was the case with the Haicheng quake. Unfortunately, large tremors are not always preceded by noticeable foreshock activity.

Squeezing or compression of the rocks adjacent to a fault might cause the ground surface to bulge upward, and thus locally tilt (Fig. 10.5). However, convincing evidence of tilt prior to earthquakes has been more difficult to acquire than might seem to be the case. The first problem is being in the right place at the right time. It is a relatively simple matter to deploy tiltmeters around an active volcano, but in the case of an earthquake how do you identify which fault is active, and how long do you monitor it or wait? Often a great deal of funds are needed to see the project through.

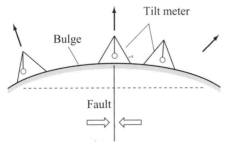

Figure 10.5   Tilt or bulging of the ground surface may occur before an earthquake because of compression and squeezing of rocks adjacent to a fault.

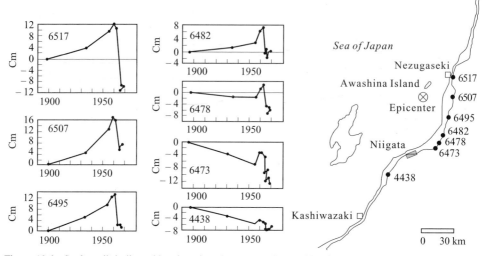

Figure 10.6   Surface tilt indicated by elevation change was detected by tidal stations along the Japanese coastline near Niigata, Japan, represents the 1964 tremor.

Few clear-cut successes of tilt as a precursor to a major tremor are known. After the fact, examination of coastal tidal gauges near Niigata, Japan, for as much as 60 years before the magnitude-7.5 1964 tremor appear to show evidence of tilt of the coastline from northeast to southwest (Fig. 10.6). However, subsequent analysis of the data has caused some to doubt the changes in elevation suggested. Certainly such changes are often small, as little as one ten-millionth of a degree of an arc tilt, and may well be due to other causes.

THE ROLE OF ROCK PROPERTIES IN CONTROLLING PRECURSORS

A seemingly fruitful approach to identifying earthquake precursors has been to study the properties of rock under stress in the laboratory. When about one-half the force that will cause a rock to fail or fracture has been applied to it, small cracks parallel to the stress direction will appear (Fig. 10.7). The rock has increased in volume and become dilatant. On the surface of Earth the increase in volume may cause a bulge and tilting of the ground. Air-filled cracks in the rock that form as a result of increase in volume change the electrical properties of the rock and also cause a drop in seismic P-wave velocity.

As stress continues to build, the cracks increase in size, allowing water to diffuse or flow into them. The presence of water causes a change of both electrical properties and seismic velocity. However, the diffusion of water into the rock weakens it, as the water creates pressure, expanding the cracks. Failure results as the cracks begin to join to form a fault, and to release the stored energy.

The dilatant-diffusion concept grew out of laboratory experiments with rock samples. It was then a natural step to apply this knowledge to active faults in the field. If the

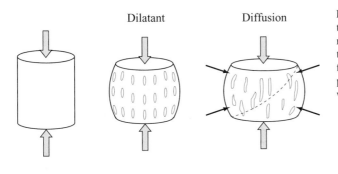

Dilatant   Diffusion

Figure 10.7   The dilatant-diffusion theory. A sequence of changes in rock under compression could lead to rock failure and earthquakes: the figure shows the dilatant or cracking phase; and the diffusion phase in which water enters the cracks.

Faulting

changes in rock properties are detected near an active fault, they may then be useful as earthquake prediction tools.

The change in electrical properties can be monitored by forcing an electric current into the ground so that it flows across a fault (Fig. 10.8). A current meter on the other side of the fault measures current flow. The difference in current between the amount flowing into the ground from the battery to that detected by the current meter is a measure of resistivity—that is, the loss of current in a material. Dry rock is highly resistant to the flow of electrical currents. On the other hand, water-saturated rock is an excellent conductor of electrical current. Recall what happens when the murderer dispatches his victim by throwing a plugged-in electrical appliance into a bathtub filled with water! Thus, resistivity falls dramatically in water-saturated rock.

If electric currents are monitored across a fault on which stress is building up, then according to the dilatant-diffusion concept, we should see two stages: first a buildup of resistivity (dilatancy), followed by a decrease in resistivity (diffusion), followed by the earthquake (Fig. 10.8).

Yet another precursor suggested from lab experiments and field records was the rate of occurrence of small earthquakes, microearthquakes. Faults with small and steady movement (creep) have concurrent microearthquakes. A rise in the rate of creep and microearthquakes may occur just before a small to moderate earthquake (Fig. 10.9). Abrupt change in creep and microearthquake rates are apparently due to an increased rate of crack formation and culminate in fault slip. A lack of microseismic activity on the other hand may suggest a buildup in stress and a locked fault. The result may be a release of stress in a large earthquake.

The dilatant-diffusion concept has also been applied to studies of seismic wave velocity. Rocks transmit shock (P-) waves and shear (S-) waves from earthquakes so that the ratio of velocities of the faster P-wave to slower S-wave is 1.75. This has been termed the Wadati ratio (Vp/Vs). Stressed rocks with air-filled cracks cause the P-wave

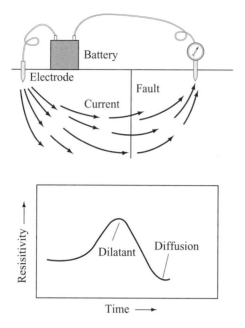

Figure 10.8   Sending an electric current across an active fault test whether or not the electrical properties of the rock have changed. According to the dilatant-diffusion concept, resistivity should rise (dilatant) and then fall (diffusion).

velocity to drop significantly, with the S-wave velocity almost unaffected. As the cracks fill with water, the P-wave velocity will rise back to more normal levels.

The phenomenon of velocity change was first noted in central Asia by Russian scientists. It was in the mid-1960s that the Russians found evidence of velocity changes. The change in the Wadati ratio might last days, weeks, or months, with P-velocity dropping below normal before climbing back up (Fig. 10.10). Shortly afterward there would be a tremor. However, there were no predictions based on this evidence, for these changes were detected by searching records after the events had occurred.

The Russian findings provided great encouragement for the international seismologic community. Personnel at the Lamont-Doherty Geological Observatory (New York) set up an experiment in the Adirondack Mountains based on the Russian re-

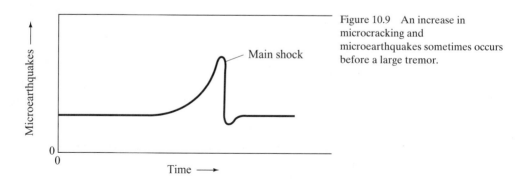

Figure 10.9   An increase in microcracking and microearthquakes sometimes occurs before a large tremor.

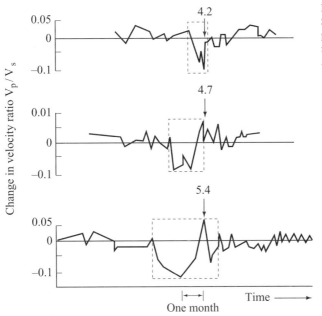

Figure 10.10   A change in the P- to S-wave velocity ratio (Wadati ratio) prior to earthquake activity was noticed in the mid-1960s in central Asia by Soviet scientists.

sults. The area for the experiment was chosen because of a swarm of microearthquakes that had occurred in 1971. Examination of records suggested that velocity changes had taken place in rocks of the area beforehand. Beginning in 1973, seismic stations were set up in the Adirondacks to monitor the Wadati ratio (Vp/Vs). By the end of July, the ratio began to drop. By August 1, the ratio had returned to normal. Based on the short interval of change in the Vp/Vs value and the area affected, a magnitude-2.5 event was predicted in a couple of days. Two days later (August 3) an earthquake of the predicted magnitude occurred. This was a stunning success. But this success was to be followed elsewhere by many cases in which little or no variation in Vp/Vs value could be detected. The principal problem is that monitoring of the kind necessary to detect velocity change is expensive. Also, as was the case with anomalous animal behavior, not all earthquakes appear to be preceded by significant or detectable velocity changes.

PARKFIELD: THE EARTHQUAKE PREDICTION EXPERIMENT

A grand experiment has been set up to detect and evaluate earthquake precursors. This involves a major and expensive effort to monitor precursors over a period of years. The site chosen for the experiment is a segment of the San Andreas fault near Parkfield, California. This is a portion of the San Andreas that is the most predictable in its behavior. The fault fails in moderate earthquakes with almost clocklike regularity, often down to even the smallest of details. Historic records show the occurrence of moderate tremors in 1857, 1881, 1901, 1922, 1934, and 1966. These were earthquakes of Richter magnitude 5.5–6.6. There is a recurrence time of about 22 years. This regularity suggested another Parkfield earthquake about 1988, perhaps as late as 1993, but unfortunately the event has not yet occurred—unfortunate because in anticipation of the tremor monitoring, efforts at prediction had been set up before 1988.

The monitoring effort at Parkfield is comprehensive. All potential precursors are observed. This includes changes in electrical resistivity and the total magnetic field as well as seismic wave velocity and microearthquake activity (Fig. 10.11). Tiltmeters have been emplaced to monitor any changes in the ground surface. Potential fault motion is monitored by distance-ranging lasers shooting light beams at targets across the fault.

Water wells are being monitored at Parkfield for evidence of change in water levels and water chemistry. As the rock around the fault cracks and fills with water, this may affect the level of water in wells in the area. Cracking of the rock could also release radon gas into the water. The rise in radon gas level was observed in the Soviet Union in the 1960s and before the 1979 Imperial Valley, California, earthquake. A relatively late addition to the precursor research effort has been the monitoring of low-frequency radio waves, which were seen to vary before the 1989 Loma Prieta tremor.

The lack of reliability of any precursors in predicting earthquakes has been disappointing and has led to dispute about the validity of the dilatant-diffusion model. The lack of success in prediction suggests that what is occurring before an earthquake is a lot more complex than we understand at present. The hope at Parkfield is that we may add

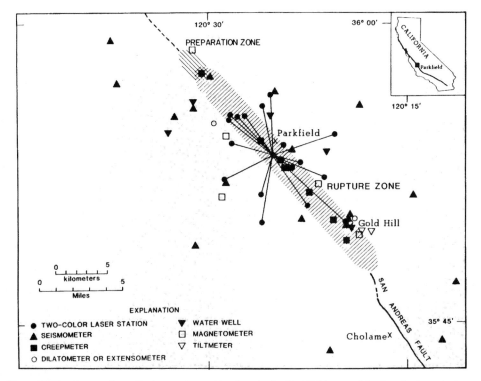

Figure 10.11   A map of monitoring equipment in place in the Parkfield area of the San Andreas fault.

new information about precursors that will be useful in leading us toward reliable prediction capability.

FOSSIL EARTHQUAKES: ROCKS TELL TALES

China has a great advantage over the rest of the world. Because of their ancient civilization and long written history, the Chinese have accumulated a wealth of earthquake information stretching back to well beyond the time before Christ. Contrast this database with that for California and the San Andreas fault. The earliest records of earthquakes in California date only to the Spanish missions of the 1700s. Two hundred years is not nearly long enough to build up a useful record of the occurrence of the large and damaging earthquakes that occur in California.

The question that can then be posed is indeed an interesting one: Do these large tremors leave behind any kind of record of their occurrence? If human records or witnesses are not available, does the ground that is ruptured and deformed in a large tremor leave evidence behind? As in all of earth science it is often useful to study the present as a key to the past. The study of modern earthquakes should provide some answers as to what to look for to find evidence of prehistoric earthquakes.

Earthquakes of magnitude ≥ 6.5 frequently leave behind several tell-tale signs that they have occurred. Such earthquakes often result in fault movements that break Earth's surface, leaving a fault scarp. Some earth scientists look for such scarps as evidence of prehistoric earthquakes. This area of study is known as *tectonic geomorphology*. Over time the fault scarps are exposed to weathering and erosion and will eventually disappear if not protected. The dry climate of the southwestern United States preserves fault scarps for several thousands of years or more, and the condition and form of the scarp can be used to give a rough idea of its age and the approximate time of the quake that produced it (Fig. 10.12).

Fault scarps are frequently buried in areas of rapid accumulation of material such as along streams or in swamps and thus are preserved for long periods of time as part of the rock record (Fig. 10.13). Key sediment layers cut by the fault in swamps contain carbon-rich material derived from the decay of plants that can be age-dated by the carbon-14 technique, giving an approximate age of the fault movement, and thus of the earthquake.

Often produced in magnitude-6.5 or larger earthquakes are sand blows, or mud volcanoes (Fig. 10.14). This occurs in the region around the fault where ground shaking of soft, water-saturated materials can be very intense. Pressure drives water-laden sand and silt or clay up through cracks in the ground to the surface where it blows into the

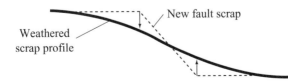

Figure 10.12   The shape of a fault scarp can be used to estimate its age and the approximate time of occurrence of the quake that produced it. The angle of slope of the fault scarp decreases with time.

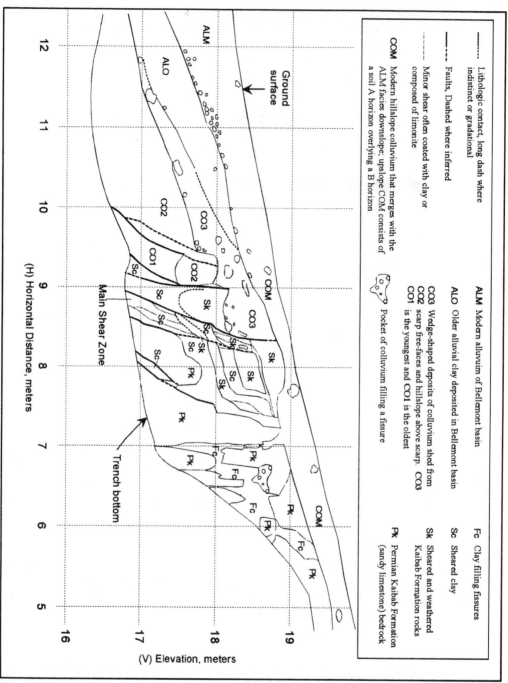

Figure 10.13   A section from a trench cut across a fault shows clearly the fault itself as well as layers adjacent to the fault used in estimating the time of occurrence of the earthquake. Layer CO3 postdates the youngest fault movement and was formed by deposition against the fault scarp. Layer CO2 has been faulted.

Figure 10.14   A sand volcano produced by ground shaking of water-saturated soils during the 1979 Imperial Valley, California, earthquake.

air, depositing a small mound of sand or mud. If these are subsequently buried and pre-served, they too can be carbon-14-dated as additional evidence of a large earthquake.

Sometimes the best or only way to see preserved and buried fault scarps and sand blows is by digging trenches across faults and looking for such evidence of a fossil earth-quake. Research of this sort is termed *paleoseismology*, or literally "ancient" seismolo-gy. This research has thus become an important tool in enlarging the earthquake database into prehistoric time.

CALIFORNIA AND THE BIG ONE

Paleoseismology is an especially important tool in southern California where the historic record of large damaging earthquakes is short, and yet the threat is real. Trenches con-structed across the San Andreas fault in southern California by Kerry Sieh of the Cali-fornia Institute of Technology have revealed a long record of large earthquakes extending back to the time of Christ. A trench across the San Andreas fault at Pallet Creek east of Los Angeles exposed the most complete record of prehistoric earthquakes known for a single fault in the world. The sediments were deposited in an ancient marsh. Evidence for a dozen large earthquakes were found (Box 10.1). The youngest event was actually historic, the great 1857 Fort Tejon earthquake. The average return time for these tremors is 145 years, but the time between events varies from 65 years to about 270 years. Working with the average number of 145 years and noting the last large event was in 1857, one might expect a recurrence of a large tremor at this locale as having a high probability by the year 2018 or any time thereafter.

EMERGENCY PREPAREDNESS

The recurrence of a Fort Tejon-sized tremor on the San Andreas fault east of Los An-geles could have a devastating effect on the Los Angeles metropolitan area. The timing would be a very important factor. If the quake were to occur at, say, 2:30 A.M., loss of life has been estimated to be 3000, whereas if it occurred at 4:30 on a weekday afternoon it

---

BOX 10.1

Fossil Earthquakes Along the San Andreas Fault

A trench across the San Andreas fault at Pallet Creek, east of Los Angeles, reveals the telltale imprint of a number of large prehistoric earthquakes. Radiocarbon dating reveals the date of each, with ± errors indicated. The most recent event is the historic 1857 Fort Tejon tremor.

Historic events	Radiocarbon dates	Prehistoric events
Z: 1857		
X: 1812	(1785 ± 32)	
	(1480 ± 15)	V
	(1346 ± 17)	T
	(1100 ± 65)	R
	(1048 ± 33)	N
	(997 ± 16)	I
	(797 ± 22)	F
	(734 ± 13)	D
	(671 ± 13)	C

could be as much as 13,000. As such, it would be the largest natural disaster ever to occur in the United States.

Efforts are going forward among emergency preparedness and government officials in California to prepare for such a disaster. Such efforts are aimed at mitigating or reducing its effects in terms of loss of lives, injuries, and financial impact. Technical advances now make possible real-time warnings of the occurrence of an earthquake. That is, within minutes of the start of a tremor, seismologists can alert emergency preparedness officials that it has occurred; they can also tell its size and where the epicenter is located. This information is invaluable to disaster-response teams and personnel at utility companies who must reroute or cut off power, water, and natural gas supplies to avoid fire and flooding.

Numerous drills in which an imaginary earthquake occurs are conducted by fire officials emergency medical personnel, police, search and rescue teams, government officials, and the California Office of Emergency Preparedness. These are often called "table top" exercises because the principal participants gather around a table and respond to the predetermined earthquake scenario.

An agreement has been worked out with Arizona at the governor's level so that emergency assistance can be provided in the case of a large tremor in southern California. As a result of the agreement, cooperative plans have been formulated to rapidly supply medical and fire support, with vehicles and personnel designated.

Building and structural failures cause much of the loss of life and economic loss in earthquakes. Mitigation or reduction of risk in this area in California consists of continued efforts to strengthen old buildings and to assure that new structures adhere to the Uniform Building Code (UBC), which is the most advanced in the world in the area of earthquake-resistant design.

The last facet of preparation for the coming large earthquake in California is public education to improve personal safety. This includes public school earthquake drills, seminars and talks before the public, and pamphlets and mailings to residents. All of these efforts are ongoing, for preparation is a mammoth task, and there are over 10 million people to prepare in one of the world's largest urban areas, Los Angeles Basin.

SUMMARY

California will suffer a great and devastating earthquake at some point in its future, of this there can be no doubt. It is the same dilemma in other earthquake-prone areas of the world, such as Mexico and Japan. But the critical question is: Where and when? Also, what do we do about it?

Two approaches have been followed to meet the threat of future earthquakes: prediction and mitigation. Prediction is the attempt to determine the time, location, and size of a future tremor. Mitigation, which is the other side of the coin, includes steps taken to reduce casualties and financial loss in the event of an earthquake.

Attempts at prediction have necessarily relied on changes that occur before the quake hits. From classical times, the Chinese in particular have considered abnormal animal behavior as an indicator of a coming tremor. Although unusual animal behavior has been widely reported, the reports have not always been gathered before a tremor. Nor have the causes of abnormal animal behavior been deduced, although hypotheses have been proposed, such as sensitivity to earthquake sounds and vibrations, or alterations in magnetic fields.

Physical changes occurring prior to earthquakes have also been well documented, but as with animal behavior, these do not seem to occur uniformly before all tremors. Physical changes include variances in earthquake wave velocity in an area where an earthquake will occur, increases in microearthquake frequency, and changes in well levels and groundwater chemistry. Variations in magnetic fields and electrical resistivity, ground level, and tilt of the surface have also been associated with changes before earthquakes.

Earthquake-wave velocity changes have been documented prior to tremors in Soviet Central Asia in the 1960s and were used to predict a small earthquake in New York State in 1973. Ground tilt was observed before the 1964 Niigata, Japan, tremor. Although there have been a few successes with animal behavior and physical precursors in predicting earthquakes, there have been many more failures.

Long-range prediction or forecasting has been somewhat more successful. Where long seismic records exist, such as in China, or rates of occurrence are high, the seismic record can be used to forecast earthquakes. One technique is to examine the historic record of seismicity and to look for geographic gaps in seismicity. Knowing that earthquakes have occurred in the seismic gap previously, it is fairly easy to predict that this gap will fill in eventually. Usually such long-range forecasts or predictions are given with a time window, and in terms of probability of occurrence for a specific area. For example, the Loma Prieta segment of the San Andreas fault had a 30 percent probability of failure in a magnitude-6.5 or larger event in 30 years.

The database in earthquake-prone regions can be extended into prehistoric time by studying clues left behind by large earthquakes: fault scarps, offset beds, and sand blows. Where such features can be dated, such as along the southern San Andreas fault, the extended record shows the recurrence of large events with an average return time of 145 years. This suggests southern California could be very close to the next Big One.

Preparations to mitigate the loss of life and property in such a large southern California tremor are well under way. This includes efforts at prediction, strengthening of buildings, preparation of response personnel, and education of the public.

KEY WORDS

Adirondacks
animal behavior
Browning
carbon-14
conditional probability
dilatant-diffusion
electrical properties
emergency preparedness
emergency response
fault scarps
forecast
ground tilt
Haicheng

lasers
Loma Prieta
long-term prediction
microearthquakes
mitigation
National Guard
New Madrid
paleoseismology
Pallet Creek
Parkfield
Perrey
precursor
prediction

prehistoric earthquakes
public education
radon
resistivity
Soviet Asia
sand blows
seismic gap
seismic velocity
short-term prediction
Tangshan
tidal pull
Wadati ratio

CHAPTER 11

What to Do Before, During, and After an Earthquake

INTRODUCTION

Although it is not possible to prevent earthquakes, it is possible to avoid or eliminate some of their dangerous effects. This is the concept of *mitigation*. Structures can be built to lessen the possibility of collapse and resulting injury or death to occupants. Fire damage can be reduced by proper precautions, such as automatic shutoffs on gas lines. These are just two examples of the efforts of government agencies toward reducing loss from earthquakes. This chapter focuses on what individuals can do to increase personal safety and reduce financial loss in earthquakes. The first step is to outline the hazards to be avoided or reduced.

EARTHQUAKE HAZARDS

The hazards created by the release of earthquake energy can be grouped into two categories: primary and secondary. Primary hazards are those directly related to ground movement. If a person is unfortunate enough to have a home built right over a fault, then when the fault moves the inevitable result is the destruction of the home. This may seem to be an unlikely situation, but southern California is laced with faults, some of which have not even been well mapped. A fault whose location is well known is the San Jacinto. Given this fact, one would think that people would avoid building structures over or close to it. Not so, for along one 5-mile stretch of the San Jacinto fault at least 20 schools lie within 2 miles of the fault (Fig. 11.1). The San Jacinto fault has historically produced more large earthquakes than any other fault in southern California.

A more widespread earthquake hazard related to the release of earthquake energy is the resultant ground shaking. Even some 200 miles away from the earthquake source, ground shaking was sufficient to bring down multistory buildings in Mexico City. The primary hazards of fault movement and ground shaking can trigger or result in secondary hazards, which may cause even greater dangers: landslides, fire, liquefaction, flooding, and tsunamis.

Consider the great fires following the earthquakes in San Francisco in 1906 and Tokyo in 1923. In these cases the havoc wrought by the fires was as great or perhaps greater than the ground shaking, both in terms of property loss and loss of life. Fire continues to be an earthquake-related problem, as, for example, at Loma Prieta (1989) and Kobe (1995).

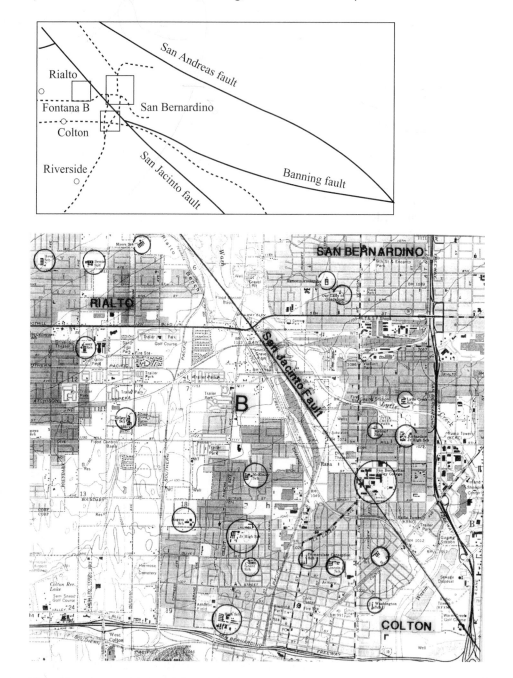

Figure 11.1   A map of the San Jacinto fault in the San Bernardino area. Shown are schools that are built close to or directly on top of the fault (circles).

Liquefaction of soil and stream deposits is a major hazard that can cause extensive damage during ground shaking. Most of the 2000 deaths in Jamaica in 1692 occurred because of the liquidlike properties of the ground, which actually swallowed up people and buildings. Any earthquake-prone areas with saturated soils or loose surface materials are at risk. This was the case at Kobe where dock facilities were severely damaged, something that literally crippled the Japanese economy for months.

Flooding can be a major hazard in areas where weak and or large dam structures exist. This is a particular problem in the Los Angeles Basin where reservoirs are held back by dams in the hills above the valleys. An example would be the Van Norman Reservoir, which sits above the San Fernando Valley north of Los Angeles and which nearly failed in the 1971 Sylmar earthquake. Had the water been released, very likely thousands in the valley below would have perished.

Landslides are an ever-present hazard when mountainous areas and earthquakes combine. Thousands died when the entire village of Yungay, Peru, was buried by soil, mud, rocks, and debris from the mountainside, which had broken loose in a large tremor. At a smaller scale, widespread damage, including road closures that cut off some communities, resulted from landsliding as a result of the 1989 Loma Prieta tremor.

Tsunamis or seismic sea waves affect only coastal areas. They are generated when faulting lifts the seafloor and the column of water above it. Such waves travel at high velocity across the ocean basins, building up a wall of water as they reach the coastline. Such a wave generated by fault movement following the 1964 Alaskan earthquake caused loss of life and damage as far away as California.

PREPARATION BEFORE AN EARTHQUAKE

Survival and overall safety in an earthquake can depend a lot on the choices made beforehand. The choice of where you decide to live or build a home is often a critical one. Obviously, one should avoid occupying a home built directly on top of, or close to, an active fault. A less obvious factor in safety is the choice of the type of material your home is built on. Generally speaking, this can make a big difference in the damage that results. Structures founded in loose soil material, whether wet or dry, are usually more heavily damaged than are those tied to bedrock, or solid rock foundation. This is because loose materials tend to amplify ground movements and cause buildings and other structures to fail. This was the case in the San Francisco Bay area in the 1989 Loma Prieta (World Series) quake. The major freeway collapse in Oakland was aided by the fact that it was founded on loose, weak substrate. The most significant damage in San Francisco was to buildings in the Marina District, which were built on bay fill material.

Location of a home in a river floodplain or other places where the soil material is likely to be saturated much of the year exposes occupants to the hazard of liquefaction during ground shaking. The soil behaves as a liquid during ground shaking, and total structural failure frequently results. Even large, well-built structures will act like ships at sea and will lurch or topple over (Fig. 11.2).

Subdivisions or municipalities developed in hilly terrain are often subject to the potential for landslide activity. Numerous examples exist of developments erected atop old landslides, which, of course, may be reactivated or covered by the deposits of new

Figure 11.2   Large apartment buildings toppled and sank as a result of liquefaction caused by the 1964 Niigata, Japan, earthquake.

ones as a result of earthquake ground shaking. After the 1971 Sylmar tremor in the Los Angeles Basin, well over a thousand small landslide movements were documented by surface mapping. Although little damage occured, a much larger tremor would result in a different story.

Another secondary earthquake hazard might exist along coastlines subjected to wave action. Hilo, Hawaii, was literally destroyed by a tsunami in 1946. Since that time much of the town has been rebuilt above the level reached by the surge of the seismic sea waves. The lesson was a graphic one: learn from the past. This is a primary weapon available to the individual in preparation for earthquake survival. If we are to reduce loss and improve safety in future earthquakes, we must learn where *not* to build.

Information derived from past tremors is available for use by individuals. Much of this information is available through governmental bodies. For instance, most states have a geological survey that publishes and makes available to the general public a wide variety of information. A very important form of information available is mapping (Fig. 11.3). Geologic maps are available showing the location of faults, types of rocks and foundation materials, and often, as well, various kinds of geologic-hazards, which include slope stability and liquefaction potential.

Many levels of government have planning bodies that may be able to provide data on places where earthquake-related hazards exist. In many states a good central location to obtain information on earthquakes or hazards in general is the state emergency services organization or the state geological survey (Appendix A). Other data sources include city and county planning boards.

It is also possible to increase personal safety in a given location in a number of ways. People are often injured or sometimes even killed by falling objects inside of buildings. To minimize such hazards do not keep heavy objects on high shelves. This would include television sets and videocassette recorders on shelf units. If such appliances must be kept high, securely fasten the shelf unit to the wall so it will not topple, and then secure the appliance to the shelf. Fires have often been started when a gas water heater is torn away from its fastenings and topples, opening up the gas line. Straps can be secured around water heaters, which are then achored to the wall (Fig. 11.4). Gas stoves can also "walk" or move during ground shaking, opening gas lines. This can be avoided by fastening stoves to the wall or floor. Not only can heavy furniture and appliances topple, causing injury, but they can move into doorways, trapping occupants inside. This is especially so for furniture such as pianos, which are on wheels and usually quite heavy.

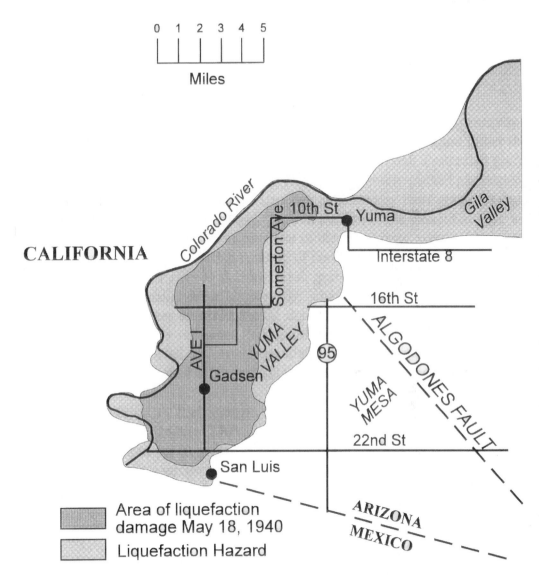

Figure 11.3   Liquefaction map of the Yuma, Arizona, area. Liquefaction occurred in this area from the 1940 Imperial Valley, California, quake. The map was prepared for the Arizona Division of Emergency Services.

Prevention of fire should be a high priority in preparing for earthquakes. Know the location of gas shut-off valves and keep a wrench nearby. Electric fires are also a possibility, so also know the location of the electric fuse box. An electric fire requires a special fire extinguisher (type C) and must not be doused with water.

Flooding can also be a problem, which can be prevented by knowing the location of the water shut-off valve. This is usually located at the water meter, the one read by meter readers periodically to determine water bills. A wrench is required to accomplish shut-off, or special shut-off tools might also be used.

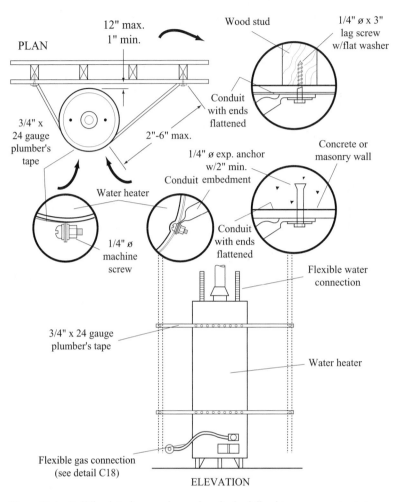

Figure 11.4 Building interiors can be made safer by following common-sense steps: Fasten water heaters to the wall; keep heavy items close to the floor; secure shelving units and bookcases so they won't topple and injure occupants.

Earthquake kits are now being sold commercially. These include critical items necessary for safety and survival after an earthquake. Usually such kits will contain a flashlight, radio, batteries for both, and a first aid kit. State emergency offices will be able to supply the names of vendors of such kits. Many of the contents, however, are common and easy to obtain, and with a little thought such a kit may be assembled by the individual. Also recommended would be stockpiling survival items for at least a 2-day period. This should include easily preserved nonrefrigerated food supplies, and bottled drinking water. In a major earthquake it may be impossible to obtain such supplies and services for up to 48 hours.

Preparation before an earthquake should include a plan for reuniting your family after the tremor in case of separation. Include out-of-state telephone numbers of friends and family to notify afterwards. It is usually impossible for outsiders to contact people in the earthquake area by phone afterwards. Often the lines will be temporarily out of service, or heavily utilized by emergency services and response personnel.

DURING AN EARTHQUAKE

Much of what should or should not be done during an earthquake is common sense. However, in the midst of ground shaking, panic frequently takes the place of rational thought. Mental preparation can help by having in mind a list of actions to be taken, and this can be the difference between survival and injury or death. The following points should be commited to memory:

1. Stay as calm as possible and do not try to move out of or into a building.
2. If you are inside a building, get under a sturdy table, or close to an inside wall.
3. If outside, find an open area and stay there. Keep away from power lines and buildings.
4. If in a vehicle, pull over and stay inside the vehicle, keeping away from bridges and overpasses.
5. High-rise buildings create special problems. Stay away from outside walls and windows, and when leaving a high-rise use the stairs.

AFTER AN EARTHQUAKE

Steps taken after an earthquake should include those that will minimize or eliminate further immediate problems that could result:

1. Prevention of fire is very important. If you detect a gas odor, open all windows and doors and shut off gas at the meter.
2. Check electric lights and water at faucets. If necessary, turn these off as well. Such steps would be necessary if flooding is in progress or if an electric fire has started or may start.
3. Do not flush toilets until sewer lines can be checked.
4. Evacuate your building if necessary. If the building you are in has been seriously damaged it could collapse as aftershocks continue to shake the ground, some of which could be almost as large as the main shock.

CASE HISTORIES: LUCK PLAYS A PART

The safest place to be in an earthquake is at home, in bed. Especially if the house is well built, single-story wood frame, and tied to bedrock by a strong foundation. If the proper preparations have been made, then a person can feel that everything possible has been done to ensure personal safety. Having said this, it is clear that circumstance and luck often play an uncomfortably strong role in survival.

Timing of an earthquake often has the largest impact on the scope of the ensuing disaster. This was clear in Lisbon on November 1, 1755, when the earthquake shocks struck on a religious holiday at a time when the churches were filled with people. The collapse of these structures killed many people who may have had a much better chance of survival had they been at home, in their beds.

California has been extremely fortunate with respect to timing of earthquakes, especially in the last several decades. The 1971 Sylmar tremor occurred just before the morning rush hour on the freeways in the Los Angeles area. Had the quake occurred a bit later during the rush to work with freeways crowded, the collapse of overpasses would have killed many more. Timing also kept the death toll low in the 1989 Loma Prieta tremor, which shook Oakland and San Francisco. The shock struck central California at about 5 P.M. on a Friday. Usually the freeways would have been clogged with vehicles. However, this Friday, because of a World Series game that was beginning in San Francisco, the freeways were relatively empty. Most people had left work early to watch the game on television. As the Series was between San Francisco and Oakland, the effect on traffic was strong. The collapse of Interstate-880 in Oakland crushed 41 people trapped in vehicles between levels of the roadway. Had this freeway been crowded with homeward-bound travelers, the death toll could easily have been several hundred (Fig. 11.5).

The Northridge, California, tremor of 1994, north of Los Angeles, also occurred at a fortuitous time. Once again, as with Sylmar in 1971, the tremor took place in the dark predawn hours before the rush hour. Again as in 1971, freeway overpasses collapsed on nearly empty roadways. Sooner or later California will run out of good luck. It is hoped that efforts underway for some years now by emergency, university, and governmental agencies will bear fruit, and that California will be even better prepared for the next damaging tremor.

Luck clearly plays a role in personal survival. Many tales of individual circumstance and its impact come from the 1964 Alaskan earthquake. The damage done to the five-story J.C. Penney department store in downtown Anchorage was witnessed by a Mrs. Tucker, who was inside the store at the time. During ground shaking, with the lights

Figure 11.5   The 1989 Loma Prieta earthquake caused collapse of the I-880 freeway in Oakland, California. Forty-one people died, trapped in cars and trucks.

out, she managed to make it from the third floor to the street entrance, with debris from the ceiling falling all about her. When she reached the street entrance, she paused for some reason that she does not understand. This saved her life, for before she could escape through the door large vertical concrete panels of the store's façade broke loose and came crashing down just outside the entrance, killing a man and woman outside. One slab crushed a parked car, reducing the vehicle to a height of 18 inches.

Just as remarkable are some of the quirky tales of survival from the gigantic seismic sea waves, or tsunamis, produced by the great Alaska tremor. Captain Cuthbert, of the crab fishing boat *Selief*, had the most remarkable tale of many to relate. The *Selief* was one boat of a fishing fleet in Kodiak Harbor, a community that relied on the sea to support itself. The first wave to come in behaved like a silent, swifly moving high tide, causing boats to sway at their moorings. When the water swept back out it left the fishing boats stranded on the harbor bottom. The big wave that came back in was a 30-foot-high wall of water that carried the 131-ton *Selief* and Captain Cuthbert over the jetty and inland. This was to be repeated a number of times until the *Selief* came to rest in the schoolhouse yard some distance from shore, with her crew unharmed.

The story at Turnagin Heights was one of the world seemingly come to an end. This area was a middle-class subdivision on Cook Inlet just south of Anchorage. The subdivision was built on unstable and soft materials. When ground shaking commenced, much of the underlying ground separated and slid into the waters, along with a number of houses. The home of the Thomas family was one of those lost by the collapsing ground. As ground shaking began, the mother and two children rushed outside only to see the entire house collapse behind them. Had they remained in the house they would have wound up with the remains of it at sea level below the bluffs some 30 feet lower.

Perhaps the clearest example the role circumstance and luck can play in survival comes from the Valdez, Alaska, area. Here ground shaking caused an enormous submarine landslide, which carried away the pier where the freighter *Chena* was unloading. The waves that were created by the landslide bobbed the *Chena* up and down like a cork on water. The ship bounced off the bottom of the bay twice and somehow remained upright. The *Chena*, her crew, and some steveadores on board helping with the unloading, survived. However, 28 steveadores, who had been unlucky enough to be on the dock when the tremor started, vanished.

A magnitude-8.1 earthquake struck Mexico City on September 19, 1985. Approximately 10,000 people died in the world's most populous urban area. Once again, survival often depended on circumstance, or good fortune. Danilo Cabrera plunged five floors in the collapse of his hotel and survived with broken ribs and cuts. Other guests of the hotel were not so fortunate.

Dr. Jose Cabanas was on the third floor of an eight-story building. After ground shaking began he managed to reach the street in front. As he did so, the building collapsed backwards, away from him. Many were killed inside. Had Dr. Cabanas left by the back entrance, he likely would have been killed as well.

Although luck cannot be controlled, certainly every possible preparation should be made to enhance personal safety. Examples have been given where circumstance determined survival. Many more examples exist where adequate and thoughtful preparation beforehand saved lives. Anything less in earthquake country is too little, too late.

SUMMARY

Earthquake survival depends in part on luck or circumstance, but also to a great degree on preparation. The hazards created by earthquakes can be reduced or mitigated by taking precautions, learned from lessons brought home from past tremors.

Siting of homes in areas where hazards are at a minimum is a critical first step. Avoidance of active faults is important to reduce damage due to fault slip and ground shaking. Safe locations can be identified from maps made available by geological surveys, state emergency management agencies, and city and county planning boards. Hazards maps, available from these agencies, identify areas of high danger from liquefaction, landslides, and flooding.

Earthquake preparation should also include making the inside of a home safe. Such steps as securing water heaters to the walls by straps and making sure that heavy items are not located on high shelves are important in reducing the potential for injury in your own home. Financial loss may be reduced by steps taken to prevent fire or flooding. Essential to this end is knowledge of where shut-off valves for gas and water are and the location of the electrical breaker panel. Additionally, appropriate fire extinguishers should be available in the home.

Preparations should be made for survival during the period immediately following the earthquake until essential services can be restored. This includes the assembly or acquisition of earthquake kits including such items as flashlights and first aid kits. Equally important would be storage of bottled water and nonperishable food supplies adequate to last at least 48 hours.

During an earthquake it is important to stay calm and to put yourself in a position where you are safe from falling objects whether inside or outside. Following the main shock, first steps include shutting off gas, water, and electricity if necessary, and evacuating your building if badly damaged, as it may collapse in aftershocks.

Many stories exist of how circumstance and simple luck combine to either enhance personal survival, or result in tragedy. Nothing can be done about this, but everything possible should be done to minimize personal risk by paying attention to factors that will certainly cause problems if not attended to e.g., knowing where utility shut-offs are located and knowing how to eliminate flooding and fire hazards that might result.

KEY WORDS

Anchorage	Hilo	San Francisco
case studies	Jamaica	secondary hazards
county planning	Kodiak	Sylmar
earthquake kits	landslides	Tokyo
emergency services	liquefaction	tsunamis
fire	Lisbon	Turnagin Heights
flooding	Mexico City	Valdez
fuse box	mitigation	Van Norman Reservoir
geological surveys	Northridge	Yungay
Hayward fault	Oakland	
hazards maps	primary hazards	

C H A P T E R   1 2

Building for Earthquake Safety

INTRODUCTION

More people are killed by the collapse of structures in earthquakes than by any other cause. This was abundantly clear in the 1989 Loma Prieta earthquake when 41 people perished in the collapse of the Interstate-880 freeway in Oakland, or in Northridge where the largest fatality count came from the collapse of an apartment building. Thus, a very important element in the reduction of the loss of lives in earthquakes is structural safety. Structural safety relies heavily on the knowledge and skills of the structural engineer and lessons learned from past earthquakes. Yet even today our knowledge remains incomplete, for there is very little information available about how buildings behave in a great earthquake.

Another problem in structural safety is deciding how safe a building should be. This is in part an economic question, for it is possible to design all buildings to be safe in the greatest of earthquakes, and to ride them out without significant damage. However, the cost of such structures may be more than is economically feasible in some situations. The question then becomes: How strong should a building be? The question has been formulated as follows: How safe should a building be? According to the Uniform Building Code (UBC), a structure should, in general, be able to:

- suffer no damage from a minor level of earthquake ground motion
- respond to a moderate level of earthquake ground motion without structural damage
- respond to a major level of earthquake ground motion having an intensity equal to the strongest experienced or forecast for that location, without collapse

This approach protects occupants first and then builds in an economically reasonable amount of survivability of the structure to minimize economic loss. This chapter focuses on structural engineering and design basics that will lead to both personal and structural survival.

THE BASICS: WHAT IS A BUILDING?

A building can be thought of as a box. At least in simplest terms that is how a structural engineer might consider it. A box has a bottom, top, and sides. The translation to a building would be foundation, roof, and walls, respectively (Fig. 12.1). The roof, and any floors, as well as beams and trusses are known as *distributing elements*, as they lie in a horizontal plane and have the function of distribution or dispersal of weight to the walls and foundations. The walls and other vertical members such as studs and columns are the *resisting elements*. They support the weight of the distributing elements and transfer the stress to the foundation. Just as important as the distributing and resisting elements are the *connectors* that tie them together.

This chapter considers the practices and materials used in building a structure and what is necessary to create a situation where collapse might be avoided. Each element will be considered in turn, beginning with the foundation.

SITE SELECTION

Arguably the most important element in any building is the foundation. Without an adequate foundation a building or other structure is certain to fail in an earthquake. The type of foundation will depend in part on the kind of structure being considered, as well as cost. In addition, foundation type may be controlled by the type of material it will be attached to or *founded* in. Foundations can be to some extent chosen or adapted to the foundation conditions. But if the foundation material is of a poor type, the cost for an adequate foundation can be very high.

The best site to build on is one of level bedrock. The foundation can be firmly attached to solid rock so that when the ground moves the structure moves with it as a unit, the same amount and in the same direction. Loose-soil sites are much less desirable, as the soil will respond by moving in different directions and amounts at the same time. In fact, loose soil has the tendency to amplify ground motion. Saturated loose soil tends to behave like a slurry during ground shaking, a time when most structures and their foundations react like tossed ships at sea. The behavior of loose material, like fill, when

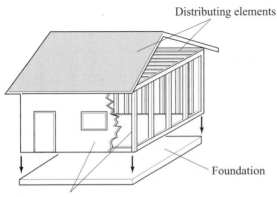

Distributing elements

Resisting elements

Foundation

Figure 12.1 The simple structural elements of a building. The roof and floors are elements that distribute or disperse weight to the walls and the foundation. The vertical members, such as the walls, resist or support the weight of the distributing elements.

it is compacted tends to be significantly better than loose soil or fill. The compression of fill by heavy equipment will strengthen it, eliminate open spaces, and help it to behave more as a unit in response to ground movement.

Problems will also result if a structure has its foundation on two or more different materials. For example, one corner of a building might be on solid bedrock while the other three corners might be on loose soil. This is a recipe for potential disaster.

Hillside locations are particularly inappropriate in earthquake-prone areas. In such an instance, not only is the site often unstable, but danger may exist from above as soil or rock material might move down onto the site. Often simply preparing such a site for a foundation can increase the instability of the slope. Poorly engineered or compacted fill or other loose material may create a potential time bomb (Fig. 12.2). In choosing a site on hillslopes, it would be wise to consult slope stability and landslide maps, which are usually available through state geological surveys and other governmental agencies.

FOUNDATIONS

The type of foundation appropriate for a structure depends on the type of building as well as the site conditions. Three basic foundation types are common in the construction industry. Perhaps the most widely used in many areas is the *stem-wall foundation*. This consists of a continuous foundation of poured concrete reinforced with vertical and horizontal steel rods, also called *reinforcing bars*, or *rebars* (Fig. 12.3). The foundation provides support for every inch of the main load-bearing outside walls of a building. Such a foundation forces the structure above it to move as a unit during an earthquake, so that damage is minimized. The stem-wall foundation is relatively inexpensive and works best when tied to solid-earth material such as bedrock.

A *mat foundation* is basically a steel rebar reinforced concrete slab resting directly on the ground (Fig. 12.4). The mat foundation is well suited for buildings on soft soil, sand, or landfill materials. The mat forms a rigid single-unit base that lessens the hazard of differential movement of soft ground by extending the foundation over most pockets of loosely packed materials. The mat foundation is also relatively inexpensive and is frequently used as a foundation for single-family dwellings and small apartment complexes in many areas of the United States.

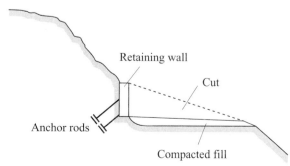

Figure 12.2   Hillside sites for structures in earthquake country require careful preparation, which would include leveling, compaction of fill, and precautions to prevent movement of material onto the site from above.

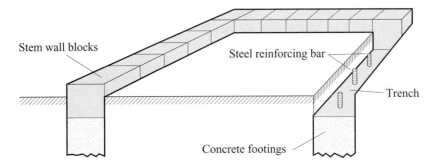

Figure 12.3 The stem-wall foundation. Such a foundation provides direct support to the walls of a structure.

Larger structures such a multistory or skyscraper buildings require more elaborate and expensive foundations (Fig. 12.5). *Pier* or *caisson pile* foundations are generally used for very tall or heavy buildings, especially those sited on clay, landfill, or generally soft ground.

The care taken with any foundation is useless unless an adequate tie exists between the foundation and the structure above it. Inadequate fastening of the structure to the foundation was not all that unusual in buildings constructed before 1930. Indeed, often private homes were simply placed on the foundation without any connections to it at all! Ground vibration will knock such buildings off the foundation or cause them to "walk off," with the result that the building will usually be a total loss (Fig. 12.6).

Present building codes enforced by local governments require that structures be tied to a foundation. There are a number of ways in which this is accomplished. Wood-frame structures can be connected to the foundation by means of bolts embedded in the concrete. J-bolts, or expansion bolts, can then be bolted to the wooden framing sill (Fig. 12.7). Such anchor bolts should be embedded a minimum of every 4 feet along the sill and within 12 inches of the ends of the sills.

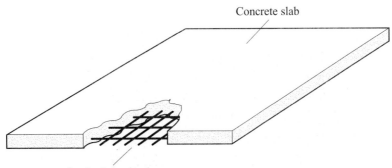

Figure 12.4 The mat foundation. This type of foundation forms a single rigid unit base and is useful where site foundation conditions vary.

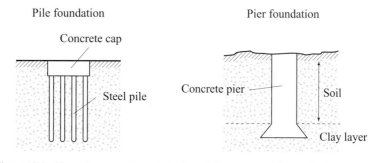

Pile foundation

Pier foundation

Figure 12.5   The caisson or pile and pier foundations are usually required to support the weight of large buildings such as skyscrapers.

WALLS

The weakest type of wall and most susceptible to failure in an earthquake is the unreinforced block or brick wall. The strong lateral forces tend to concentrate in the weakest link, which is the mortar between the bricks, bursting the wall apart. The best support is to add vertical steel rods so as to tie the brick together.

Vertical elements of a wall are termed *columns*. They can be of virtually any material, from wood to concrete or steel. The vertical elements in wood-frame buildings are called *studs*, and they are important weight-bearing and weight-distribution elements. Their stability during earthquake ground movement is dependent on two factors: connections and bracing.

The manner in which columns are connected to horizontal wall elements such as the top sill and bottom sill is critical if a building is to resist collapse during a tremor. The wood-frame building columns can be connected by bolts, or steel angles of various designs (Fig. 12.8). Steel columns in large buildings can be welded or bolted, or even both, to the horizontal steel members of the frame. During the Northridge, California, tremor of 1994, connectors in steel-frame buildings were weakened by cracking, a totally unexpected response. Had these weakened connections not been discovered, it is conceivable that they, and the buildings containing them, would have failed in a subsequent earthquake.

Figure 12.6   Failure of a home in the Santa Cruz, California, area that was not well tied to its foundation.

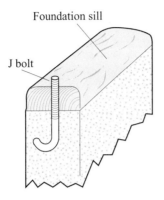

Foundation sill

J bolt

Figure 12.7 Connection of a wood-frame wall to a foundation is done by means of J-bolts embedded in the stem-wall foundation.

Bracing of walls is important to ensure survival of a building in an earthquake. There are two philosophies with respect to bracing. These are the strengthening of the wall to resist the motion imparted to the building, and the opposite approach, which is to absorb the ground motion by response of the building frame.

Large multistory buildings are usually designed with frame-action bracing (Fig. 12.9). Flexible connectors between vertical and horizontal elements allow the building to give and absorb the ground motion, without failure. A very tall skyscraper might sway at the top as much as a few feet laterally without collapse. However, the trick is to make the connectors flexible and strong. Apparently, more remains to be learned, as was found out from Northridge.

Walls can be strengthened by adding sheets of material between columns. This approach is termed *shear-wall bracing*, and the shear walls can consist of anything from plywood to concrete slabs (Fig. 12.10). The shear wall provides a resistant element

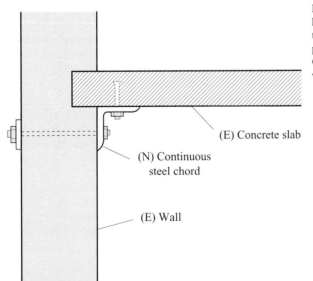

(E) Concrete slab

(N) Continuous steel chord

(E) Wall

Figure 12.8 Connection of horizontal elements such as floors to the walls can be critical in preventing structural collapse. Good connections can be made with bolts or steel angles.

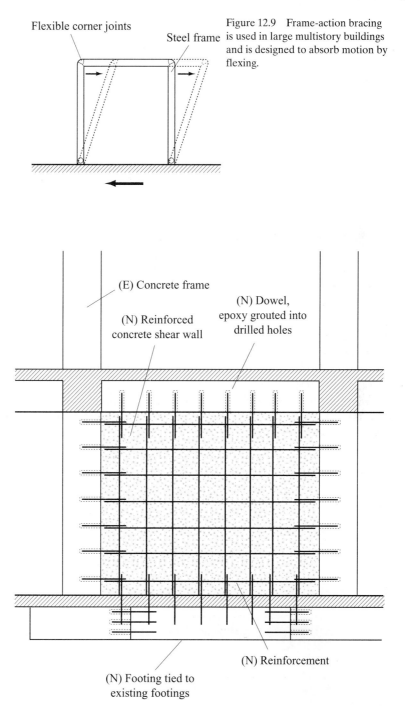

Flexible corner joints

Steel frame

Figure 12.9   Frame-action bracing is used in large multistory buildings and is designed to absorb motion by flexing.

(E) Concrete frame

(N) Dowel, epoxy grouted into drilled holes

(N) Reinforced concrete shear wall

(N) Reinforcement

(N) Footing tied to existing footings

Figure 12.10   Shear-wall bracing is designed to resist rotation and collapse of a building by preventing movement of elements such as columns. Cross bracing has the same purpose and is a less expensive approach.

preventing the columns from rotating and collapsing. Nowhere was the role of shear-wall bracing more clearly illustrated than in the damage suffered by the new Olive View Veterans Hospital in the 1971 Sylmar (California) earthquake. The building had originally been designed to have concrete shear walls between columns on the ground floor. A last-minute change eliminated the shear walls to let in more light to the first-floor areas. As a result of the lack of shear walls, the columns failed in the tremor, and although the main building did not collapse, the damage was not repairable, and the building had to be demolished (Fig. 12.11).

Another approach to strengthening walls is by *cross bracing*. This is a much simpler and less expensive technique and therefore one usually employed in smaller structures, such as private homes. Cross bracing is the insertion of angled beams between columns, to resist column rotation. The beam could consist of wood in a wood-frame structure but could also be of other materials.

Connections between walls and floors also may help to strengthen a structure. Poor floor-to-wall connections are a common problem in masonry buildings with wood floors. The heavy walls tend to pull away from the floor or roof and sever connections, resulting in collapse. This can be avoided by use of steel connections (Fig. 12.8).

THE ROLE OF DESIGN IN SAFETY

A carefully designed building is the best survival insurance available in earthquake-prone areas. The role of a knowledgeable architect is invaluable. Such a professional would recognize the importance of shape in increasing the survivability of a structure. The more boxlike a building, the safer it will be. Buildings with wings, or different levels in different parts of a structure, pose a potential risk. Unless care is taken in design and construction practices, the connection between the main part of a building and a wing may focus destructive ground movements and amplify their effects. An example can be seen with the elevator shaft and stairwell wings of the Olive View Veterans Hospital in

Figure 12.11   Damage to the brand-new Olive View Veterans Hospital in 1971 could have been prevented by the addition of shear walls between the first-floor columns.

Figure 12.12   The connection between the main building and the elevator stairwell wing of the Olive View Veterans Hospital proved weak and failed during the 1971 Sylmar, California, earthquake.

the 1971 Sylmar tremor (Fig. 12.12). This section actually separated and pulled away from the main building, toppling over. This, of course, effectively traps people above the ground level.

A building of varying height may cause problems because the various levels will cause the building to vibrate at two different frequencies, rather than moving as a unit, thus amplifying the destructive effects of ground motion. Hence, the simpler the shape and the more boxlike the structure, the fewer problems that are likely to result from a tremor.

Yet another problem created by inappropriate design practices is the *soft-story approach*. An example of the soft-story approach would be a two-story home where the street level consists of a garage. The open space of the garage and lack of interior walls create a weaker level within the home. The horizontal shearing forces concentrate at this level because of the added weight of the floor above. Because of inertia and mass, the upper floor tends not to move, whereas the lower floor moves sideways with the ground, creating a rotating force on the garage walls, and leading to collapse (Fig. 12.13). The ground floor of the Olive View Veterans Hospital was a soft story because of a combination of recessed walls and a lack of shear-wall bracing on the outer columns.

ORNAMENTAL DESIGN AND CHIMNEYS

Heavy vertical elements not well attached to the building are often the first to fail by collapse during ground shaking. This includes cornices, parapets, false building fronts, and chimneys.

Chimneys behave like inverted pendulums and can topple in moderate tremors (Fig. 12.14). The greatest hazard from a collapsing chimney would be if it fell inward, into the home. The most hazardous type of chimney is one of unreinforced masonry, similar to those on many older buildings. Because of the higher price of masonry construction, newer chimneys in many parts of the United States consist of the less expensive single-unit metal-flue type, encased in an insulated wood frame. Not only are these chimneys lighter and thus more stable, but the uniformly strong steel pipe tends to behave as a single flexible unit during ground movement. However, no chimney is safe unless properly connected to the building structure. This can be achieved with masonry chimneys by

Figure 12.13   This private home in Sylmar, California, had a first-floor garage. The garage served as a weak level or soft story, which led to the collapse of the home in the 1971 tremor.

utilizing steel straps attached to both the wooden or block frame of the house and to the chimney itself (Fig. 12.15).

During the first part of the twentieth century in the United States it was common practice to attach ornamental masonry walls, often including heavy parapets or cornices, to the front of wooden buildings to increase the impression of viewing from the street a more expensive structure. Many of these buildings still exist across the country, especially in large urban areas such as San Francisco and Los Angeles. During an earthquake there is a tendency for these false fronts to behave very much like chimneys and to col-

Figure 12.14   Chimneys can be particularly hazardous architectural elements in a home, unless securely attached to the dwelling. New steel flue chimneys are both safer and cheaper to install.

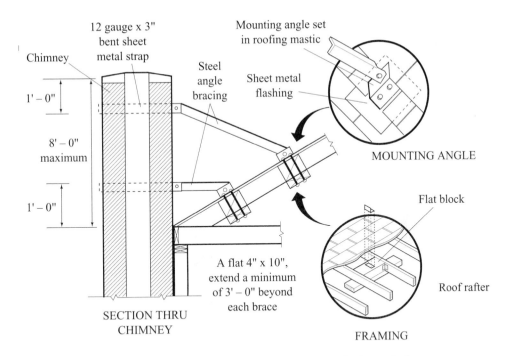

12 gauge x 3"
bent sheet
metal strap

Chimney

Mounting angle set
in roofing mastic

Steel
angle
bracing

Sheet metal
flashing

1' – 0"

8' – 0"
maximum

1' – 0"

MOUNTING ANGLE

Flat block

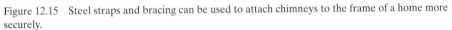

A flat 4" x 10",
extend a minimum
of 3' – 0" beyond
each brace

Roof rafter

SECTION THRU
CHIMNEY

FRAMING

Figure 12.15   Steel straps and bracing can be used to attach chimneys to the frame of a home more securely.

lapse into the streets, crushing pedestrians and vehicles. This is precisely what happened in Santa Cruz in the older downtown shopping district (Fig. 12.16). As with chimneys, there are ways of reinforcing false-masonry storefronts to prevent, or reduce, the possibility of collapse.

Parapets can be lowered to remove the weight to a lower level and then strengthened by inserting steel belts or straps, more securely tying the parapet to the building. False fronts can be anchored to the building framework by steel ties spaced at least within 1.5 feet of one another. This assumes that the basic building is structurally sound,

Figure 12.16   False fronts in older structures are often improperly attached to the rest of the building. Heavy masonry fronts are especially likely to pull away from wooden-frame structures during a tremor.

for in an earthquake the collapse of a false front or veneer can bring down the building itself with it.

MOBILE HOMES

Mobile homes risk extensive damage in a tremor because of inadequate connections between the mobile home and its foundation. Many states have no particular requirements in their building codes in this respect. A frequent approach is simply to place the mobile home on stilts or piers, allowing the weight of the mobile home to keep it in place and stable on these supports. The 1971 Sylmar tremor shook numerous mobile homes in the area off their supports, causing extensive damage and injuries to the occupants.

The safety of mobile homes can be increased with little extra cost because the foundation area is easily accessible. Cross bracing with steel cables anchored to the ground is sufficient to resist separation of the mobile home from its supports (Fig. 12.17). Strong support piers are also important. These should ideally be of reinforced concrete block with adequate footings.

Both structural engineering and architectural design have come a long way since 1900 and have benefitted from lessons learned in past earthquakes. Today it is possible to design buildings that will be much safer in earthquakes than those of a century ago. The most serious problems remaining in building safety are enforcement of building codes and the safety of older buildings that are not up to modern standards.

SUMMARY

The application of design knowledge in building construction from structural engineers and architects is essential for building safety in earthquake country. Much has been learned about the behavior of foundations, walls, floors, and ornamental elements from damage incurred in past tremors.

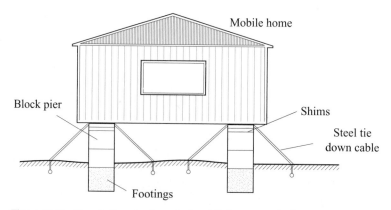

Figure 12.17   Foundation attachment of mobile homes can be increased by the use of cross-braced steel cables. The foundation piers can also be strengthened by steel rebar.

Critical to the role of the foundation in an earthquake is site selection. Site selection will determine the type of foundation needed as well as the likely survival of a structure in a tremor. Sites that might be a problem include hillsides and those with soft, loose materials, such as fill. Sites not on bedrock would be well served by mat foundations for smaller buildings and pier or caisson pile foundations for larger commercial or business structures.

All elements of a building should be well connected. This includes the foundation to the rest of the building, as well as the floor and roof to walls. The walls should be adequately strengthened to resist rotation and consequent collapse of the structure. This can be achieved through several different kinds of bracing: frame action, shear wall, or cross bracing. A special case is the mobile home, which is usually placed on piers or pilings with no secure attachment. Stability can be increased inexpensively by the use of steel tie-down cables.

Simple overall design of the building will improve survivability of the structure from ground shaking. The more boxlike the structure, the less likely that failure will occur. Anything violating this principle, such as weak or soft-story designs, should be avoided.

Ornamental design features and chimneys are often a hazard in earthquakes. False-masonry building fronts, cornices, and parapets are prone to collapse, especially with older structures where they are often unreinforced or poorly attached to the rest of the building. Masonry chimneys also pose a threat, especially if they collapse into a building. Chimneys can be reinforced and securely fastened to a building using bolts and steel straps.

KEY WORDS

architect
caisson foundation
chimneys
columns
compressed fill
connectors
cross bracing

distributing elements
expansion bolts
false fronts
frame-action bracing
mat foundation
mobile homes
parapets

rebar
resisting elements
shear walls
soft story
stem-wall foundation
structural engineer

Government Emergency Services and Geoscience Organizations

FEDERAL

Federal Emergency Management Agency (FEMA)
500 C Street S.W.
Washington, DC 20472

National Geophysical Data Center (NOAA)
NOAA/NESDIS
325 Broadway
Boulder, CO 80303
Tel. (303) 497-6215; Fax (303) 497-6513

U.S. Geological Survey
National Center
12201 Sunrise Valley Drive
Reston, VA 20191
Tel. (703) 648-4000; Fax (703) 648-4454

STATE ORGANIZATIONS

Alabama Geological Survey, 420 Hackberry Lane, P.O. Box O, Tuscaloosa, AL 35486-9780; Tel. (205) 349-2861

Alaska Division of Emergency Services, P.O. Box 5750, Fort Richardson, AK 99505-5750; Tel. (907) 474-7147; Fax (907) 479-4779

Alaska Earthquake Information Center, Geophysical Institute, University of Alaska Fairbanks, 903 Kotukuk Dr., Box 747320, Fairbanks, AK 99775-7320

Alaska State Geological Survey, 794 University Ave., Suite 200, Fairbanks, AK 99709-3645; Tel. (907) 451-5001; Fax (907) 451-5050

Arizona Division of Emergency Management, 5636 E. McDowell Rd., Phoenix, AZ 85008; Tel. (602) 231-6245; Fax (602) 231-6231

Arizona Earthquake Information Center, Box 4099, Northern Arizona University, Flagstaff, AZ 86011-4099; Tel. (520) 523-7197; Fax (520) 523-9220

Arizona Geological Survey, 416 West Congress St., Suite 100, Tucson, AZ 85701; Tel. (520) 770-3500; Fax (520) 770-3505

Arkansas Geological Commission, Vardelle Parham Geology Center, 3815 West Roosevelt Rd., Little Rock, ARK 72204; Tel. (501) 296-1877; Fax (501) 663-7360

California Office of Emergency Services, Metro Center, 101 8th St., Suite 152, Oakland, CA 94607; Tel. (510) 540-2713; Fax (510) 540-3581

California Division of Mines and Geology, 801 K St., MS 14-33, Sacramento, CA 95814-3532

Colorado Office of Emergency Management, Division of Local Government, 15075 South Golden Rd., Camp George West, Golden, CO 80401-3979; Tel. (303) 273-1779; Fax (303) 273-1795

Colorado Geological Survey, 1313 Sherman St., Room 715, Denver, CO 80203; Tel. (303) 866-2611; Fax (303) 866-2461

Connecticut Geological and Natural History Survey, Dept. of Environmental Protection, Natural Resources Center, 79 Elm St., Hartford, CT 06106-5127; Tel. (860) 424-3540; Fax (860) 424-4058

Delaware Geological Survey, University of Delaware, Delaware Geological Survey Building, Newark, DE 19716-7501; Tel. (302) 831-2833; Fax (302) 831-3579

District of Columbia, Geologist of Washington, DC, University of District of Columbia, Dept. of Biological and Environmental Science, 4200 Connecticut Ave., N.W., MB 44-04, Washington, DC 20008-1154; Tel. (202) 274-5886; Fax (202) 274-5952

Florida Geological Survey, Florida Dept. of Environmental Protection, 903 West Tennessee St., Tallahassee, FL 32304-7700; Tel. (904) 488-4191; Fax (904) 488-8086

Georgia Geologic Survey, Georgia Department of Natural Resources, Environmental Protection Division, 19 M.L. King Jr. Drive, Room 400, Atlanta, GA 30334-9004; Tel. (404) 656-3214; Fax (404) 651-9425

Hawaii Geological Survey, Division of Water/Land Development, Department of Land and Natural Resources, P.O. Box 373, Honolulu, HI 96809; Tel. (808) 587-0230; Fax (808) 587-0219

Hawaii State Civil Defense, 3949 Diamond Head Rd., Honolulu, HI 96816-4495; Tel. (808) 734-2161; Fax (808) 737-4150

Idaho Geological Survey, University of Idaho, Morrill Hall, Room 332, Moscow, ID 83843; Tel. (208) 885-7991; Fax (208) 885-5826

Idaho Military Division, Bureau of Disaster Services, P.O. Box 83720, 650 West State St., Boise, ID 83720-0023; Tel. (208) 334-3460; Fax (208) 334-2322

Illinois State Geological Survey, 615 East Peabody Drive, Champaign, IL 61820-6964; Tel. (217) 333-5111; Fax (217) 244-7004

Indiana Geological Survey, Indiana University, 611 North Walnut Grove, Bloomington, IN 47405; Tel. (812) 855-5067; Fax (812) 855-2862

Iowa Department of Natural Resources, Geological Survey Bureau, 109 Trowbridge Hall, Iowa City, IA 52242-1319; Tel. (319) 335-1575; Fax (319) 335-2754

Kansas Geological Survey, University of Kansas, 1930 Constant Avenue, Campus West, Lawrence, KS 66047; Tel. (913) 864-3965; Fax (913) 864-5317

Kentucky Geological Survey, University of Kentucky, 228 Mining and Minerals Resources Building, Lexington, KY 40506-0107; Tel. (606) 257-5500; Fax (606) 257-1147

Louisiana Geological Survey, Louisiana State University, P.O. Box G, Baton Rouge, LA 70893; Tel. (504) 388-5320; Fax (504) 388-5328

Maine Geological Survey, Department of Conservation, Natural Resources and Mapping Center, State House Station, Number 22, Augusta, ME 04333; Tel. (207) 287-2801; Fax (207) 287-2353

Maryland Geological Survey, Maryland Department of Natural Resources, 2300 St. Paul St., Baltimore, MD 21218-5210; Tel. (410) 554-5500; Fax (410) 554-5502

Office of State Geologists, Commonwealth of Massachusetts, 100 Cambridge St., 20th Floor, Boston, MA 02202; Tel. (617) 727-5830; Fax (617) 727-2754

Michigan Department of Environmental Quality, Geological Survey Division, P.O. Box 30256, Lansing, MI 48909; Tel. (517) 334-6923; Fax (517) 334-6038

Minnesota Geological Survey, University of Minnesota, Twin Cities, 2642 University Ave. W., St. Paul, MN 55114-1057; Tel. (612) 627-4780; Fax (612) 627-4778

Mississippi Office of Geology, Mississippi Department of Environmental Quality, P.O. Box 20307, 2380 Highway 80 West, Jackson, MS 39289-1307; Tel. (601) 961-5500; Fax (601) 354-7151

Missouri Department of Natural Resources, Geological Survey Program, P.O. Box 250, Buehler Building/111 Fairgrounds Rd., Rolla, MO 65401-0250; Tel. (314) 368-2101; Fax (314) 368-2111

Montana Bureau of Mines and Geology, Montana Tech of the University of Montana, 1300 West Park St., Butte, MT 59701-8997

Montana Division of Disaster and Emergency Services, P.O. Box 4789, Helena, MT 59604-4789; Tel. (406) 444-6911; Fax (406) 444-6965

Nebraska Geological Survey, University of Nebraska-Lincoln, Conservation and Survey Division, 901 North 17th St., 113 Nebraska Hall, Lincoln, NE 68588-0517; Tel. (402) 472-3471; Fax (402) 472-2410

Nevada Bureau of Mines and Geology, University of Nevada, Reno, Mail Stop 178, Reno, NV 89557-0088; Tel. (702) 784-6691; Fax (702) 784-1709

Nevada Division of Emergency Management, Capitol Complex, 2525 South Carson St., Carson City, NV 89710; Tel. (702) 687-4240; Fax (702) 687-6788

New Hampshire Geological Survey, Department of Environmental Services; Tel. (603) 271-3406; Fax (603) 271-6588

New Jersey Geological Survey, CN 427, Trenton, NJ 08625; Tel. (609) 292-1185; Fax (609) 633-1004

New Mexico Bureau of Mines and Mineral Resources, New Mexico Institute of Mining and Technology, Campus Station, Socorro, NM 87801; Tel. (505) 835-5420; Fax (505) 835-6333

New Mexico Emergency Management Bureau, P.O. Box 1628, Santa Fe, NM 87504-1628; Tel. (605) 827-9222; Fax (605) 827-3456

New York State Geological Survey, 3140 Cultural Education Center, Albany, NY 12230; Tel. (518) 474-5816; Fax (518) 473-8496

North Carolina Geological Survey, Department of Environment, Health, and Natural Resources, Box 27687, 512 North Salisbury St., Raleigh, NC 27611; Tel. (919) 733-2423; Fax (919) 733-0900

North Dakota Geological Survey, 600 East Blvd., Bismarck, ND 58505-0840; Tel. (701) 328-9700; Fax (701) 328-9898

Ohio Department of Natural Resources, Division of Geological Survey, 4383 Fountain Square Drive, Columbus, OH 43224-1362; Tel. (614) 265-6988; Fax (614) 268-3669

Oklahoma Geological Survey, University of Oklahoma, 100 East Boyd, Energy Center, Room N-131, Norman, OK 73019-0628; Tel. (405) 325-3031; Fax (405) 325-7069

Oregon Department of Geology and Mineral Industries, 800 NE Oregon St., Suite 965, Portland, OR 97232-2162; Tel. (503) 731-4100; Fax (503) 731-4066

Oregon Emergency Management, 595 Cottage St. N.E., Salem, OR 97310; Tel. (503) 378-2911; Fax (503) 588-1378

Pennsylvania Bureau of Topographic and Geologic Survey, DCNR-Pennsylvania Geological Survey, P.O. Box 8453, Harrisburg, PA 17105-8453; Tel. (717) 787-2169; Fax (717) 783-7267

Geological Survey of Rhode Island, Office of the State Geologist, University of Rhode Island, 315 Green Hall, Kingston, RI 02881; Tel. (401) 874-2265; Fax (401) 874-2190

South Carolina Geological Survey, Department of Natural Resources, 5 Geology Rd., Columbia, SC SC 29210-9998; Tel. (803) 896-7700; Fax (803) 896-7695

South Dakota Geological Survey, University of South Dakota, Dept. of Water and Natural Resources, Science Center, Vermillion, SD 57069-2390; Tel. (605) 677-5227; Fax (605) 677-5895

Tennessee: Center for Earthquake Research and Information (CERI), Memphis State University, Memphis, TN 38152; Tel. (901) 678-2007; Fax (901) 678-4734

Tennessee Division of Geology, Department of Environment and Conservation, Division of Geology, 13th Floor, L & C Tower, 401 Church St., Nashville, TN 37243-0445; Tel. (615) 532-1500; Fax (615) 532-0231

Bureau of Economic Geology, University of Texas, Austin, University Station, Box X, 10100 Burnett Rd., Austin, TX 78713-7508; Tel. (512) 471-1534; Fax (512) 471-0140

Utah Geological Survey, Utah Department of Natural Resources, 1594 West North Temple, Suite 3110, P.O. Box 146100, Salt Lake City, UT 84114-6100; Tel. (801) 537-3300; Fax (801) 537-3300

Utah Division of Comprehensive Emergency Management, State Office Building, Room 1110, Salt Lake City, UT 84114; Tel. (801) 538-3758; Fax (801) 538-3772

Vermont Geological Survey, Agency of Natural Resources, 103 South Main St., Laundry Building, Waterbury, VT 05671-0301; Tel. (802) 241-3608; Fax (802) 241-3273

Virginia Division of Mineral Resources, Department of Mines, Minerals, and Energy, Fontaine Research Park, 900 Natural Resource Drive, P.O. Box 3667, Charlottesville, VA 22901; Tel. (804) 293-5121; Fax (804) 293-2239

Washington Department of Natural Resources, Geology/Earth Resources, P.O. Box 47007, Olympia, WA 98504-7007; Tel. (360) 902-1450; Fax (360) 902-1785

Washington Divison of Emergency Management, P.O. Box 48346, 4220 E. Martin Way, Olympia, WA 98504-8346; Tel. (206) 459-9191; Fax (206) 923-4991

West Virginia Geological and Economic Survey, P.O. Box 879, Morgantown, WV 26507-0879; Tel. (304) 594-2331; Fax (304) 594-2575

Wisconsin Geological and Natural History Survey, University of Wisconsin-Extension, 3817 Mineral Point Rd., Madison, WI 53705; Tel. (608) 262-1705; Fax (608) 262-8086

Wyoming State Geological Survey, P.O. Box 3008, University Station, Laramie, WY 82071; Tel. (307) 766-2286; Fax (307) 766-2605

Wyoming Emergency Management Agency, P.O. Box 1709, Cheyenne, WY 82003-1709; Tel. (307) 777-4900; Fax (307) 635-6017

Computer-Based Earthquake Information

The following list of addresses is not meant to be an exhaustive one but merely a starting point. Especially useful for beginners will be the University of Washington address (Washington seismosurfing), which is a catalogue of other addresses.

LISTS OF RECENT EARTHQUAKES CAN BE OBTAINED BY THE FINGER COMMAND; TYPE:

finger quake@gldfs.cr.usgs.gov	A worldwide list of larger earthquakes
finger quake@scec.gps.caltech.edu	Southern California earthquake information
finger quake@andreas.wr.usgs.gov	Northern California earthquake information
finger quake@geophys.washington.edu	Northwest U.S. earthquake information
finger quake@fm.gi.alaska.edu	Alaskan earthquake information
finger quake@seismo.unr.edu	Nevada earthquake information
finger quake@eqinfo.seis.utah.edu	Utah and Yellowstone earthquake information
finger quake@slueas.slu.edu	Central U.S. earthquake information

EARTHQUAKE INFORMATION ON THE WORLD WIDE WEB (WWW). AFTER HTTP:// TYPE:

wwwneic.cr.usgs.gov/neis/bulletin/bulletin.html	Latest worldwide earthquake information
www.ceri.memphis.edu/	Center for EarthquakeResearch/Information
www.scecdc.scec.org	Southern California earthquakes/research
quake.wr.usgs.gov	Northern California earthquakes/research
www.iris.edu	Incorporated Research Institutions for Seismology—Global Research Programs
eri.u-tokyo.ac.jp/	Source for Japanese Earthquake data
cdidc.org	Test Ban Treaty Data—Worldwide Explosions
geophys.washington.edu/seismosurfing.html	A general directory of internet earthquake addressess—a good place to begin
seismology.harvard.edu	Harvard seismology information
pgc.nrcan.gc.ca	Pacific Geoscience Centre—earthquakes
tlacaelel.igeofcu.unam.mx	University of Mexico (National)—earthquakes
seismo.ethz.ch	Surfing for European earthquake information
vishnu.glg.nau.edu/aeic/aeic.html	Arizona earthquakes
quake.geo.berkeley.edu/cnss	Authoritative composite earthquake catlogue— Council of the National Seismic System

APPENDIX C

Suggested Readings

Guide: G = readable by the general public; no previous background necessary; T = technical; professional and/or college level; T/G = partially useful or suitable to both general and professional/technical readers.

BOOKS (ARRANGED CHRONOLOGICALLY)

1927   **Founders of Seismology**, Davison, C., Cambridge Press, Cambridge, England (G)

1954   **The Birth and Development of the Geological Sciences**, Adams, F. D., Dover, New York (G)

1958   **Elementary Seismology**, Richter, C., W. H. Freeman and Co., San Francisco (T)

1971   **The San Francisco Earthquake**, Thomas G., and Witts, M. M., Dell, New York (G)

1971   **Earthquake Country**, Iacopi, R., Lane Book Co., San Francisco (G)

1977   **Earthquakes**, DeNevi, D., Celestial Arts, Millbrae, CA (G)

1982   **Planet Earth: Earthquake**, Walker, B., Time-Life, Alexandria, VA (G)

1984   **Terra Non Firma**, Gere, J., and Shah, H., Stanford Alumni Association, Stanford, CA (G)

1985   **Evaluating Earthquake Hazards in the Los Angeles Region—An Earth Science Perspective**, Ziony, J., United States Government Printing Office, Washington, DC (T)

1990   **Peace of Mind in Earthquake Country**, Yanev, P., Chronicle Books, San Francisco (G)

1990   **The Solid Earth: An Introduction to Global Geophysics**, Cambridge Press, Cambridge, England (T)

1992   **Inside the Earth**, Bolt, B., W. H. Freeman and Co., San Francisco (G)

1993   **Earthquakes**, Bolt, B., W. H. Freeman and Co., San Francisco (G)

1993   **Responses to Iben Browning's Prediction of a 1990 New Madrid, Missouri, Earthquake**, Spence, W., Herrmann, R., Johnston, A., and Reagor, G., U.S.G.S. Circular 1083, United States Government Printing Office, Washington, DC (G)

1993   **The Practice of Earthquake Hazard Assessment**, McGuire, R., U.S.G.S. Federal Center, Denver (contact: Engdahl) (T)

1994   **Images of the 1994 Los Angeles Earthquake**, Los Angeles Times Staff, Los Angeles Times, Los Angeles (G)

1995   **Reducing Earthquake Losses**, Herdman, R., U.S. Government Printing Office, Washington, DC (T/G)

1996   **Active Tectonics: Earthquakes, Uplift, and Landscape**, Keller, E., and Pinter, N., Prentice-Hall, Upper Saddle River, N.J. (T/G)

1997   **The Geology of Earthquakes**, Yeats, R., Sieh, K., and Allen, C., Oxford Univ. Press, New York, (T)

PERIODICALS

Bulletin of the Seismological Society of America: 201 Plaza Professional Building, El Cerrito, California 94530 (T)

Journal of Geophysical Research-Seismology: American Geophysical Union, 2000 Florida Avenue, N.W., Washington, DC 20009 (T)

Journal of Seismology: Kluwer Academic Publishers, 101 Philips Dr., Assinippi Park, Norwell, Massachusetts 02061 (T)

PERIODICAL ARTICLES ON EARTHQUAKES (ARRANGED IN CHRONOLOGICAL ORDER)

A. National Geographic

1986   Earthquake in Mexico, Boraiko, A., v. 169, p. 655–675 (G)

1988   The Day the World Ended at Kourion, Soren, D., v. 174, no. 1, p. 30–53 (G)

1990   Earthquake: Prelude to the Big One?, v. 177, no. 5, p. 76–105 (G)

1995   Living with California's Faults, Gore, R., v. 187, no. 4, p. 2-35 (G)

B. Newsweek

1989   The Lessons of the Bay Quake, October 30, 1989, p. 22(G)

1994   After the Quake, Adler, J., January 31, 1994, p. 25–33 (G)

1995   Lessons of Kobe, Begley, S., January 30, 1995, p. 24–27 (G)

C. Time

1994   State of Shock, Gibbs, N., v. 143, no. 5, p. 26–37 (G)

1995   Killer Quake: When Kobe Died, Van Biema, D., v. 145, no. 4, p. 24–33(G)

ILLUSTRATION CREDITS

Cover–lower illustration (freeway ramp): Steinbrugge Collection, Earthquake Engineering Research Center, University of California, Berkeley.

Figure 1.1 Michele Brumbaugh.

Figure 1.2 Teledyne-Geotech.

Figure 1.4 Caso, (1978), *The Aztecs, People of the Sun*. University of Oklahoma Press, p. 32, by permission of Fondo de Cultura Economica U.S.A., San Diego, CA.

Figure 1.5 *Philosophical Transactions* (1761), v. LI, Table XIII, p. 585.

Figure 1.6 Steinbrugge Collection, Earthquake Engineering Research Center, University of California (Berkeley).

Figure 1.7 Richter, (1958), *Elementary Seismology*. W. H. Freeman, San Francisco. Fig. 28–7, p. 482.

Figure 2.1 Davison, (1927), *The Founders of Seismology*. Cambridge University Press, Cambridge, Fig. 6, p. 122.

Figure 2.3 Borchardt, (editor) (1975), Studies for Seismic Zonations of the San Francisco Bay Region. U.S.G.S. Professional Paper 941-A, Fig. 2.8.

Figure 2.4 Espinosa, A.(1977) *Earthquake Information Bulletin*, March-April 1977, v. 9, no. 2, Fig. 3, p. 9.

Figure 2.5 Munoz, and Udias, (1988), Evaluation of Damage and Source Parameters of the Malaga Earthquake of 1680, Fig. 2, p. 212. In *Historical Seismograms and Earthquakes of the World*, Lee, Meyers, and Shimazaki, (editors). Academic Press, New York.

Box 2.1 After U. S. G. S. Earthquake Report Form.

Figure 2.6 Hansen, (1965), Effects of the Earthquake of March 27, 1964, at Anchorage, Alaska. U.S.G.S. Professional Paper 542-A, Fig. 2.6.

Figure 2.11 *The Engineer*, v. 33, 1877.

Box 2.5 *Nature*, 1886–1887.

Figure 2.23 Mykkeltveit, Ringdal, Kvaerna, and Alewine, (1990), Application of Regional Arrays in Seismic Verification Research. *Bulletin of the Seismological Society of America*, v. 80, Fig. 2, p. 1780.

Figure 3.1 Art Sylvester, University of California, Santa Barbara.

Box 3.2 *Earthquake Information Bulletin* (1973), July–Aug. 1973, vol. 5, no. 4, Fig. 1, p. 6.

Figure 3.8 Wald, Helmberger, and Heaton (1991), Rupture Model of the 1989 Loma Prieta Earthquake from the Inversion of Strong Motion and Broad Band Teleseismic Data. *Bulletin of the Seismological Society of America*, v. 81, no. 5, p. 1564.

Figure 3.9 Susong, Janecke, and Bruhn (1990), Structure of a Fault Segment Boundary in the Lost River Fault Zone, Idaho, and Possible Effect on the 1983 Borah Peak Earthquake. *Bulletin of the Seismological Society of America*, v. 80, no. 1, pp. 64–65.

Figure 3.10 Brumbaugh (1984), Compressive Strains Generated by Normal Faulting. *Geology*, v. 12, p. 494, and Brumbaugh and Dresser (1976), Exposed Step in a Laramide Thrust Fault, Southwest Montana. *American Association of Petroleum Geologists Bulletin*, v. 60, no. 12, p. 2147, ©1984, Geological Society of America.

Box 3.3 Pelton, Meissner, and Smith (1984), Eyewitness Account of Normal Surface Faulting. *Bulletin of the Seismological Society of America*, v. 74, no. 3, p. 1086; Wallace (1984), Eyewitness Account of Surface Faulting During the Earthquake of 28 October 1983, Borah Peak, Idaho. *Bulletin of the Seismological Society of America*, v. 74, no. 3, p. 1091.

Figure 3.17 Marshall, Stein, and Thatcher (1991), Faulting Geometry and Slip from Co-Seismic Elevation Changes. *Bulletin of the Seismological Society of America*, v. 81, no. 5, p. 1667.

Figure 3.18 Steinbrugge Collection, Earthquake Engineering Research Center, University of California, Berkeley.

Figure 3.19 After Earthquakes and Volcanoes, *Northridge Special Issue* (1994), v. 25, no. 1, p. 6.

Figure 4.6 Pinar, Honkura, and Kikuchi (1994), Rupture Process of the 1992 Erzincan Earthquake and Its Implication for Seismotectonics in Eastern Turkey. *Geophysical Research Letters*, v. 21, no. 18, p. 1972.

Figure 4.8 Wadati (1928), Shallow and Deep Earthquakes. *The Geophysical Magazine*, v. 1, Fig. 2, p. 175.

Figure 4.11 Johnston (1990), An Earthquake Strength Scale for the Media and the Public. *Earthquakes and Volcanoes*, v. 22, no. 5, pp. 214–216.

Figure 5.2 Honda (1932), On the Mechanism and the Types of the Seismograms of Shallow Earthquakes. *Geophysical Magazine*, v. 5, Fig. 4, p. 73.

Figure 5.6 Brumbaugh (1979), Classical Focal Mechanism Techniques. *Geophysical Surveys*, v. 3, Fig. 4, p. 303.

Figure 5.8 Brumbaugh (1979), Classical Focal Mechanism Techniques. *Geophysical Surveys*, v. 3, Fig. 11, p. 310.

Figure 5.9 Rogers and Lee (1976), Seismic Study of Earthquakes in the Lake Mead, Nevada-Arizona Region. *Bulletin of the Seismological Society of America*, v. 66, no. 5, Fig. 11, p. 1675.

Figure 5.15 Pinar, Honkura, and Kikuchi (1996), A Rupture Model for the 1967 Mudurnu Valley, Turkey, Earthquake and Its Implications for Seismotectonics in the Western Part of the North Anatolian Fault Zone. *Geophysical Research Letters*, v. 23, no. 1, Fig. 3, p. 31.

Figure 5.17 Source model solution (Mw=5.3) in figure is from: Lay, Ritsema, Ammon, and Wallace (1994), Rapid Source Mechanism Analysis for the April 29, 1993, Cataract Creek (Mw=5.3) Northern Arizona Earthquake. *Bulletin of the Seismological Society of America*, v. 84, no. 2, Fig. 2, p. 453.

Figure 5.18 Pinar, Honkura, and Kikuchi (1996), A Rupture Model for the 1967 Mudurnu Valley, Turkey, Earthquake and Its Implications for Seismotectonics in the Western Part of the North Anatolian Fault Zone. *Geophysical Research Letters*, v. 23, no. 1, Fig. 2, p. 30.

Figure 5.19 Wald, Helmberger, and Heaton (1991), Loma Prieta Earthquake from the Inversion of Strong Motion and Broadband Teleseismic Data. *Bulletin of the Seismological Society of America*, v. 81, no. 5, Fig. 13, p. 1564.

Figure 5.20 Stein, Okal, and Weins (1988), The Application of Modern Techniques to Analysis of Historical Earthquakes, *Historical Seismograms and Earthquakes of the World*, Meyers and Schimazaki (editors), Academic Press, Fig. 2, p. 89.

Figure 6.6 Raff and Mason (1961), Magnetic Survey off the West Coast of North America 40N Latitude to 52N Latitude. *Geological Society of America Bulletin*, v. 72, no. 8, Fig. 1, p. 1268, ©1961, Geological Society of America.

Figure 6.8 Brumbaugh (1979), Classical Focal Mechanism Techniques. *Geophysical Surveys*, v. 3, Fig. 25, p. 326.

Figure 6.9 Brumbaugh (1979), Classical Focal Mechanism Techniques. *Geophysical Surveys*, v. 3, Fig. 1, p. 298.

Figure 6.10 Brumbaugh (1979), Classical Focal Mechanism Techniques. *Geophysical Surveys*, v. 3, Fig. 24, p. 324.

Figure 6.11 Brumbaugh (1979), Classical Focal Mechanism Techniques. *Geophysical Surveys*, v. 3, Fig. 26, p. 327.

Box 6.2 Benioff (1955), Seismic Evidence for Crustal Structure and Tectonic Activity. In *Crust of the Earth* (Poldervaart, A., editor). *Geological Society of America Special Paper* 62, Fig. 8, p. 68, ©1955, Geological Society of America.

Box 7.1 Genkin, Gorbunov, Kazansky, Lanev, Filimonova and Yakovlev (1987), Ore Mineraliza- tion, Fig. 1.53, p. 206; and Kozlovsky (editor), Fig. 1.1, p. 2, in *Superdeep Well of the Kola Penin- sula*. Academic Press, New York.

Figure 7.12 Hirakawa, (1981), Three Dimensional Seismic Structures Beneath Southwest Japan, The Subducting Phillipine Sea Plate. *Tectonophysics*, v. 79, pp. 1–44.

Figure 7.13 Steeples, and Iyer, (1976), Low Velocity Zone Under Long Valley as Determined from Teleseismic Events. *Journal of Geophysical Research*, v. 81, no. 5, Fig. 5b, p. 857.

Figure 8.1 Soren (1988), The Day the World Ended at Kourion. *National Geographic*, v. 174, no. 1, pp. 38–39.

Figure 8.7 Link (1960), Exploring the Drowned City of Port Royal. *National Geographic*, v. 117, no. 2, pp. 166–167.

Figure 8.8 Reid (1914), The Lisbon Earthquake of November 1, 1755. *Bulletin of the Seismologi- cal Society of America*, v. 4, no. 2, p. 56, *after* Fig. 1.

Figure 8.9 Birmingham (1993), *A Concise History of Portugal*, plate 20, p. 74. Cambridge Univer- sity Press, Cambridge.

Figure 8.11 Dubois and Smith (1980), The 1887 Earthquake in San Bernardino Valley, Sonora: Historical Accounts and Intensity Patterns in Arizona. *Arizona Geological Survey, Special Paper no. 3.*

Figure 8.12 Steinbrugge Collection, Earthquake Engineering Research Center, University of Cal- ifornia, Berkeley.

Figure 8.13 Steinbrugge Collection, Earthquake Engineering Research Center, University of Cal- ifornia, Berkeley.

Figure 8.15 The Library of Congress.

Figure 8.16 Pflaker and Savage (1970), Mechanism of the Chilean Earthquakes of May 21 and 22, 1960. *Geological Society of America Bulletin*, v. 81, no. 4, *after* Fig. 1, p. 1003, ©1970, Geologi- cal Society of America.

Figure 8.17 Pflaker and Savage (1970), Mechanism of the Chilean Earthquakes of May 21 and 22, 1960. *Geological Society of America Bulletin*, v. 81, no. 4, *after* Fig. 2, p. 1005, ©1970, Geologi- cal Society of America.

Figure 8.18 Steinbrugge Collection, Earthquake Engineering Research Center, University of Cal- ifornia, Berkeley.

Figure 8.20 Steinbrugge Collection, Earthquake Engineering Research Center, University of Cal- ifornia, Berkeley.

Figure 8.21 Steinbrugge Collection, Earthquake Engineering Research Center, University of Cal- ifornia, Berkeley.

Figure 8.22 Steinbrugge Collection, Earthquake Engineering Research Center, University of Cal- ifornia, Berkeley.

Figure 9.2 Bausch and Brumbaugh (1996), Yuma Community Earthquake Hazard Evaluation. Arizona Division of Emergency Management, cover illustration.

Figure 9.3 Natural Hazard Photographs: National Oceanic and Atmospheric Administration, World Data Center A for Solid Earth Geophysics, National Geophysical Data Center, 82- EHB-02, B79J15-020.

Figure 9.11 Richter (1958), *Elementary Seismology*. W. H. Freeman & Co., San Francisco, Fig. 28–21, p. 511.

Figure 9.12 Steinbrugge Collection, Earthquake Engineering Research Center, University of Cal- ifornia, Berkeley. Volcano Monitoring at the U. S. G. S. Hawaiian Volcano Obsevatory.

Figure 9.14 Heliker, Griggs, Takahishi, and Wright, (1986), *Earthquakes & Volcanoes*, Jan.–Feb. 1986, v. 18, no. 1, p. 32.

Figure 9.15 Tilling, (1976), The 7.2 Magnitude Earthquake, November 1975, Island of Hawaii. *Earthquake Information Bulletin*, v. 18, no. 6, pp. 5–13.

Figure 9.19 Stauder (1977) Microearthquake Array Studies of the Seismicity in Southeast Missouri. *Earthquake Information Bulletin*, v. 9, no. 1, Fig. F, p. 12.

Figure 10.1 Tasch, in *Essentials of Geology*, 5th ed., Fig. 15.22, p. 329, Prentice-Hall, New Jersey.

Box 10.1 Data from Sieh, Stuiver, and Brullinger (1989), A More Precise Chronology of Earthquakes Produced by the San Andreas Fault in Southern California. *Journal of Geophysical Research*, v. 94, no. B1, pp. 603–623. Data from table 3, p. 614.

Figure 10.2 United States Geological Survey (1989), Lessons Learned from the Loma Prieta Earthquake of October 17, 1989. *U.S.G.S. Circular 1045*.

Figure 10.6 Fujii and Nakane (1996), Reevaluation of Anomalous Vertical Crustal Movement Associated with the 1964 Niigata Japanese Earthquake, Case 24. In *Earthquake Prediction State of the Art*, Birkhauser, Basel, Wyss and Dmouska (editors).

Figure 10.10 Kisslinger (1969), *Earthquake Prediction*, p. 41; Semyenov, A. N. (1969), Variations in the Travel-Time of Transverse and Longitudinal Waves Before Violent Earthquakes. *Izvestia Academy Science* U.S.S.R. Physics of the Solid Earth (English ed.) v. 4, pp. 245–248.

Figure 10.11 United States Geological Survey (1986), *Earthquakes and Volcanoes*, v. 18, no. 4, p. 149.

Figure 10.13 Pearthree, Vincent, Brazier, and Hendricks (1996), Plio-Quaternary Faulting and Seismic Hazard in the Flagstaff Area, Northern Arizona. *Arizona Geological Survey Bulletin*, v. 200, Fig. 8, p. 11.

Figure 10.14 Natural Hazard Photographs (1982), National Atmospheric and Oceanic Administration, World Data Center A for Solid Earth Geophysics, National Geophysical Data Center, 82-EHB-02, B79J15-018.

Figure 11.2 Steinbrugge Collection, Earthquake Engineering Research Center, University of California, Berkeley.

Figure 11.3 Bausch and Brumbaugh (1996), Yuma Community Earthquake Hazard Evaluation. Arizona Division of Emergency Management, Fig. 4, p. 20.

Figure 11.4 Federal Emergency Management Agency (1994), Reducing the Risks of Non-Structural Earthquake Damage: FEMA 74, 3rd ed., Fig. U19a, p. 35.

Figure 11.5 Steinbrugge Collection, Earthquake Engineering Research Center, University of California, Berkeley.

Figure 12.6 Steinbrugge Collection, Earthquake Engineering Research Center, University of California, Berkeley.

Figure 12.8 Federal Emergency Management Agency (1992), *Handbook for Seismic Rehabilitation of Existing Buildings*. FEMA 172, Fig. 3.5.4.3, p. 58.

Figure 12.10 Federal Emergency Management Agency (1992), *Handbook for Seismic Rehabilitation of Existing Buildings*. FEMA 172, Fig. 3.1.2.2c, p. 26.

Figure 12.11 Steinbrugge Collection, Earthquake Engineering Research Center, University of California, Berkeley.

Figure 12.12 Steinbrugge Collection, Earthquake Engineering Research Center, University of California, Berkeley.

Figure 12.13 Larry Reddinger

Figure 12.14 Steinbrugge Collection, Earthquake Engineering Research Center, University of California, Berkeley.

Figure 12.14 Federal Emergency Management Agency (1994), *Reducing the Risks of Non-Structural Earthquake Damage*: FEMA 74, Fig. U38, p. 44.

Figure 12.16 Steinbrugge Collection, Earthquake Engineering Research Center, University of California, Berkeley.

GLOSSARY

aftershocks smaller-magnitude earthquakes that occur after the main shock and in the region close to the mainshock.

amplitude the peak height of a wave crest or lowest point of a wave trough.

asperity a rough spot on a fault surface.

asthenosphere the layer below the lithosphere at depths from about 100 kilometers beneath Earth's surface to a depth of about 200 kilometers. The rock material that makes up the asthenosphere is very warm and thought to be soft or weak.

attenuation a loss in wave height or amplitude of earthquake waves.

auxiliary plane the nodal plane in a fault-plane solution that is perpendicular to the fault plane.

basalt a rock type formed by the crystallization of a lava flow that is rich in minerals containing silicon, oxygen, iron, and magnesium.

Basin and Range a region of the western United States dominated by a thin crust, abundant normal faults, and an alternating mountain/valley topography.

bedrock solid and generally unaltered rock material, often buried beneath weathered debris.

blind fault a fault that occurs within the subsurface and that has no surface expression/exposure.

body waves earthquake waves that pass through Earth's interior rather than traveling along the surface. Examples would be both the primary (P) and secondary (S) waves.

broadband having a wide range of sensitivity. Usually refers in earthquake studies to a special seismometer design that has a broad range of relatively high sensitivity to ground motion.

bulk modulus the resistance that an elastic material has to a change in volume.

clipping the square-wave response seen on seismograms that are due to the ground motion exceeding the dynamic range of the seismograph system.

compression force that is directed such that it may result in shortening or reduction in volume of solid material.

conduction the physical process whereby heat is transferred in a solid from particle to particle, while the particles themselves remain in a fixed position.

convection the physical process whereby heat is transferred by movement of the heated material.

continental drift the concept that Earth's continents move on the planet's surface by plowing through the rocks of the ocean floor. A concept associated with Alfred Wegener.

core the central region of Earth that lies directly adjacent to and below the mantle.

cornice the relatively small wall-like projection at the top of a building, usually for ornamental purposes.

creep slow slip or movement along a fault plane without accompanying earthquakes.

critical angle the angle of approach of a wave to a boundary that results in refraction of the wave parallel to and below the boundary.

cross bracing the *x*-pattern bracing between vertical structural elements of a building wall that has the effect of strengthening the wall. The braces can be of different materials (e.g., wood).

damping a progressive reduction in a cyclic phenomenon. It can refer to the loss of wave energy or amplitude due to friction, or the reduction in the amount of swing of a seismometer boom.

density the mass per unit of volume of substances, such as rock.

diffusion the intermingling of two substances, often used to describe the dispersion of a fluid through a porous solid. An example would be water *diffuses through* porous rock.

dilatant the increased state of volume of a solid.

dip the angle of inclination of a surface, such as a fault plane, from the horizontal.

dip-slip fault a fault in which the slip or movement of the fault blocks is in a direction parallel to the dip (inclination) of the fault surface.

distributing elements horizontal elements (e.g., floors) in a building structure that distribute the weight downward.

double-couple an arrangement of forces consisting of two perpendicular single-couples.

elastic the physical property that solid materials have of recovering their initial shape or volume after the release of a deforming force.

elastic rebound theory the theory that includes a description of the role of elastic behavior of rocks under strain before, during, and after an earthquake.

epicenter the point on Earth's surface that lies above the earthquake focus.

false fronts ornamental fronts of buildings; often of heavier materials, such as brick, than the rest of the structure.

fault a fracture along which motion of the two sides of the fracture has occurred parallel to the fracture surface.

fault-plane solution a stereographic plot of earthquake-wave first motions that defines the orientation of the fault plane involved, as well as the type of faulting.

first motion a term used to describe the first ground motion of an arriving earthquake wave as seen on a seismogram.

focus the point within Earth where an earthquake fault begins to rupture. Also known as the *hypocenter*.

forecast a less precise estimate than prediction of the timing of a future tremor.

foreshock a smaller earthquake that precedes the main shock.

frequency number of cycles or oscillations per second. One hertz = one cycle per second.

fossil earthquake a prehistoric earthquake that has left evidence of its occurrence in the rock record.

graben a linear block of downfaulted rock bounded by normal faults along each side.

hertz see *frequency*.

hot spot a center of volcanism located within a tectonic plate.

hypocenter see *focus*.

ice age a term used to describe a time of global glaciation. Frequently applied to the most recent glacial epoch within the last million years.

impulsive sudden or rapid. The term is applied to the sharp, clear onset of earthquake waves as seen on seismograms.

inertia the tendency for a body to remain unchanged in its state of motion. If at rest, to remain at rest; if in motion, to continue in motion. The larger the mass, the stronger the inertia.

intensity a measure of the level of ground shaking based on damage reports and observer-felt reports.

Intermountain Seismic Belt a geographic belt of earthquake activity stretching north-south approximately from Canada to southern Utah.

isoseismal a contour separating areas on a map of different levels of ground shaking.

kilometer a metric system unit of distance measurement. One kilometer is equal to .625 mile.

lava molten rock or magma that has reached the surface of Earth.

liquefaction the fluidlike behavior of water-saturated soil or other loose and fine-grained material.

lithosphere the outermost rigid, brittle rock layer of Earth that includes the crust and uppermost mantle.

loess fine-grained glacially derived sediment.

Love waves surface earthquake waves that force material in their path to shear or move sideways perpendicular to the direction of the wavepath.

low velocity zone a zone of decreased seismic wave velocity found in the upper mantle.

magma molten rock material located within Earth.

magnitude an instrumental measure of the size of an earthquake.

mantle the largest rock layer of Earth's interior that lies between the crust and the core.

meizoseismal zone the area on Earth's surface that contains the strongest ground shaking and damage.

microearthquake the smallest earthquakes, usually considered to be those with a magnitude of less than 3.0.

mitigation a term applied to planning used to reduce casualties and property loss from earthquakes.

Mohorovicic discontinuity the boundary between the crust and the mantle. The Moho lies at depths ranging from 5 kilometers under the oceans to about 60 kilometers under young mountain ranges.

moment a measure of earthquake size that can be related to the area of the fault surface that ruptured and to the properties of the rock in which the fault is located.

moment magnitude earthquake size that is estimated by using the seismic moment.

mud volcanoes a small topographic feature produced by the expulsion of water and fine-grained material during severe ground shaking from earthquakes. See *sand blows*.

nodal plane a plane of no earthquake-wave first motion, separating quadrants of opposite senses of motion in a fault-plane solution.

noise term used to refer to all ground movements not related to earthquakes or explosions (e.g., from wind).

normal fault a fault that forms due to a stretching or lengthening of Earth's crust. Slip is parallel to the dip of the fault surface, with relative movement of the upper block down and the lower block up.

ocean ridge a seafloor mountain range formed by volcanic activity at a plate tectonic boundary.

origin time the time at which seismic waves originate at the focus.

paleoseismology the study of prehistoric earthquake activity.

period the difference in the arrival time of successive wave crests of a wave train.

plate the lithospheric plates of plate tectonics.

precursor a change that occurs before an event. Physical precursors such as tilt of Earth's surface sometimes occur before earthquakes.

perovskite a dense oxide mineral compound that contains calcium and titanium.

polar reversal the periodic switch of Earth's north and south magnetic poles.

precursor a term applied to the physical changes that might take place before an earthquake.

prediction the ability to determine the time, size, and location of an earthquake before the fact.

P-wave the primary or first-arriving seismic wave from an earthquake. P-waves force the material in their paths to compress and expand in the direction of wave travel.

Rayleigh wave an earthquake wave that travels along Earth's surface. Material in the path of a Rayleigh wave is forced to move in a vertical elliptical pattern.

recurrence interval the time interval between earthquakes of a given magnitude.

resisting elements the vertical structural elements of a building.

resistivity term used to describe the resistance to the flow of current in a material.

resonance the relatively strong vibrations of an oscillating system such as a pendulum.

reverse fault a fault in which the block above the fault plane moves up relative to the block below. The movement is in the direction of inclination, or dip, of the fault plane.

rift a tear or split in the rocks of Earth's crust, resulting from extension of the crust and marked at the surface by normal faults and grabens.

rigidity the elastic property of solids measured by the resistance of the solid to a change in shape.

salt dome a domelike mass of salt that has been emplaced in the surrounding rock material by moving or flowing upward.

sand blows see *mud volcanoes*.

scarp a topographic cliff. A fault scarp is that part of the fault plane exposed at Earth's surface.

seafloor spreading motion of the seafloor caused by movement of the lithospheric plate containing the oceanic crust.

seiche oscillation of water in a bay or lake often caused by earthquake ground motion.

seismic gap an area in which there is reduced magnitude or frequency of earthquakes when compared to the adjacent or surrounding area.

seismic tomography the study of Earth's interior in three dimensions using images created from analyses of variations in velocity of seismic waves.

seismic wave an elastic wave in rock that is produced by an explosion or earthquake.

seismicity a term that refers to some aspect of earthquake activity, such as "the seismicity of southern California" or "seismicity has decreased through time."

seismograph an instrument that produces a chronologic record of ground movement during an earthquake.

seismogram a paper record of ground movements that is produced by a seismograph system.

seismology the study of earthquakes and associated phenomena.

seismometer the part of a seismograph that actually responds to or senses ground motion.

seismoscope an instrument that detects ground motion but does not produce a permanent record of it.

shadow zone the area on Earth's surface that does not receive waves from an earthquake because of interference or blocking action by Earth's core.

single-couple two oppositely directed parallel forces separated by a specified perpendicular distance.

slip the relative motion of one block over a fault surface compared to the other block.

soft story a weak level or story within a building susceptible to collapse during ground shaking.

sonar a depth-sounding device using sound waves to gauge underwater depths.

source in earthquake seismology, a term used to refer to the area of release of earthquake energy.

stem wall a type of building foundation that consists of concrete and block footings beneath the outside walls.

stereographic the representation of three-dimensional information in a two-dimensional diagram.

strain the temporary or permanent response to stress. The response might be a change in position, a rotation, or a deformation that could be a change in shape or volume, or any combination of these.

stress the intensity of force as measured by the amount of force over a designated area; for example, kilograms (static) per square centimeter.

strike the compass direction of a line. The line in question could be formed by the intersection of a fault plane or joint plane with Earth's surface.

strike-slip fault a fault with the slip directed in the direction of the fault strike—that is, horizontal slip with respect to Earth's surface.

subduction zone a zone of contact between two converging tectonic plates in which one plate is forced down into Earth.

surface waves earthquake waves traveling along Earth's surface.

S-wave an earthquake wave that travels through Earth's interior and that forces particles of material in its path to move from side-to-side, perpendicular to the wavepath.

tectonics large-scale deformation of the rocks of Earth's lithosphere.

teleseism a distant earthquake.

thrust fault a reverse fault with a low angle of inclination of the fault plane.

transform fault a strike-slip fault found at plate tectonic boundaries that facilitates the horizontal slip of adjacent plates past one another.

travel-time the time in seconds or minutes that it takes an earthquake wave to travel between two points (e.g., the earthquake focus and a seismic station).

travel time curve a graphic plot of travel time versus distance for earthquake waves. The curves can be used to determine velocity of waves, earthquake origin time, and earthquake location.

trench a linear depression in Earth's surface. A plate tectonic trench marks the surface boundary between two plates at a subduction zone.

tsunami a seismic sea wave created by vertical movement of the seafloor resulting from fault slip during an earthquake.

Universal Coordinated Time standard 24-hour time used by seismologists and astronomers. Time is calibrated to that kept at Paris, France, where midnight is 0 hour.

volcano an opening or vent in Earth's crust that allows magma to escape in the form of lava or airborne fiery particles.

Wadati ratio the ratio of the P-wave velocity to the S-wave velocity, often about 1.75.

wavelength the distance measured along the wavepath between two wave crests or wave troughs.

Index